BUSINESS ISSUES, COMPETITION AND ENTREPRENEURSHIP

CRAMMING AND THIRD PARTY BILLING ON MOBILE PHONES

ANALYSES AND PERSPECTIVES

Business Issues, Competition and Entrepreneurship

Additional books in this series can be found on Nova's website under the Series tab.

Additional e-books in this series can be found on Nova's website under the e-book tab.

BUSINESS ISSUES, COMPETITION AND ENTREPRENEURSHIP

CRAMMING AND THIRD PARTY BILLING ON MOBILE PHONES

ANALYSES AND PERSPECTIVES

CHRISTOPHER RYE
EDITOR

nova publishers
New York

For permission to use material from this book please contact us:
Telephone 631-231-7269; Fax 631-231-8175
Web Site: http://www.novapublishers.com

Library of Congress Cataloging-in-Publication Data

ISBN: 978-1-63463-121-1

Published by Nova Science Publishers, Inc. † New York

CONTENTS

PREFACE

For several decades, phone companies have allowed third-party vendors to charge consumers on their phone bills for goods and services unrelated to phone service, such as photo storage, voicemail, and faxes. This practice began with landline phone bills and continued on wireless phone bills as consumer use of mobile phones increased. Throughout this period, the industry has assured the public that its self-regulatory system is effective at protecting consumers from fraudulent third-party billing on their phone bills. However, this Committee's 2010-2011 review of third-party billing practices on landline phones showed that widespread unauthorized charges – known as "cramming" – had been placed on phone bills and had likely cost consumers billions of dollars over the preceding decades. In light of these findings, and emerging reports of cramming in the wireless context, the Committee subsequently began reviewing third-party billing practices on wireless phone bills. This book provides and analyses and perspectives on cramming and third party billing on mobile phones.

Chapter 1 – This is a report, issued by the Senate Committee on Commerce, Science, and Transportation on Cramming on Mobile Phone Bills: A Report on Wireless Billing Practices, dated July 30, 2014.

Chapter 2 – This is a report by Federal Trade Commission on Mobile Cramming: An FTC Staff, dated July 2014.

Chapter 3 – This is a Statement of John R. Thune, Ranking Member, Senate Committee on Commerce, Science, and Transportation. Hearing on "Cramming on Wireless Phone Bills: A Review of Consumer Protection Practices and Gaps" before the Senate Committee on Commerce, Science, and Transportation, dated presented July 30, 2014

Chapter 4 – This is a Statement of Terrell McSweeny, Commissioner, Federal Trade Commission. Hearing on "Cramming on Wireless Phone Bills:

A Review of Consumer Protection Practices and Gaps" before the Senate Committee on Commerce, Science, and Transportation, dated July 30, 2014.

Chapter 5 – This is a Testimony of William H. Sorrell, Attorney General, State of Vermont. Hearing on "Cramming on Wireless Phone Bills: A Review of Consumer Protection Practices and Gaps" before the Senate Committee on Commerce, Science, and Transportation, dated July 30, 2014.

Chapter 6 – This is a Testimony of Travis LeBlanc, Acting Chief, Enforcement Bureau, Federal Communications Commission. Hearing on "Cramming on Wireless Phone Bills: A Review of Consumer Protection Practices and Gaps" before the Senate Committee on Commerce, Science, and Transportation, dated July 30, 2014.

Chapter 7 – This is a Testimony of Michael F. Altschul, Senior Vice President and General Counsel, CTIA - The Wireless Association. Hearing on "Cramming on Wireless Phone Bills: A Review of Consumer Protection Practices and Gaps" before the Senate Committee on Commerce, Science, and Transportation, dated July 30, 2014.

Chapter 8 – This is a consumer guidance document by Federal Communications Commission on Cramming: Unauthorized, Misleading, or Deceptive Charges Placed on Your Telephone Bill, dated January 2013.

In: Cramming and Third Party Billing ... ISBN: 978-1-63463-121-1
Editor: Christopher Rye

Chapter 1

CRAMMING ON MOBILE PHONE BILLS: A REPORT ON WIRELESS BILLING PRACTICES*

Office of Oversight and Investigations

EXECUTIVE SUMMARY

For several decades, phone companies have allowed third-party vendors to charge consumers on their phone bills for goods and services unrelated to phone service, such as photo storage, voicemail, and faxes. This practice began with landline phone bills and continued on wireless phone bills as consumer use of mobile phones increased. Throughout this period, the industry has assured the public that its self-regulatory system is effective at protecting consumers from fraudulent third-party billing on their phone bills.

However, this Committee's 2010-2011 review of third-party billing practices on landline phones showed that widespread unauthorized charges – known as "cramming" – had been placed on phone bills and had likely cost consumers billions of dollars over the preceding decades.

In light of these findings, and emerging reports of cramming in the wireless context, the Committee subsequently began reviewing third-party billing practices on wireless phone bills.

* This is an edited, reformatted and augmented version of a Majority Staff report, issued by the Senate Committee on Commerce, Science, and Transportation, July 30, 2014.

This inquiry focused largely on third-party vendor charges placed through a system known as the premium short message service, or "PSMS," which involves use of text messaging charged to consumers at a higher rate than standard text messaging. These types of charges had been the focus of mounting reports of abuses. Products charged to consumers through the PSMS system generally have involved digital goods used on mobile phones, such as ringtones and cellphone wallpaper, or for services such as subscriptions to periodic text message content sent to the subscriber on subjects such as horoscopes or celebrity gossip.

To assess the nature and scale of wireless cramming, the Committee's majority staff reviewed narrative and documentary information provided by the four major wireless carriers, entities known as "billing aggregators" that serve as middlemen between vendors and carriers in the billing process, and other sources.

Unfortunately, the information reviewed by the Committee shows that, just as in the landline context, cramming on wireless phone bills has been widespread and has caused consumers substantial harm. Specifically, this report finds:

- Third-party billing on wireless phone bills has been a billion dollar industry that has yielded tremendous revenues for carriers. AT&T, Sprint, T-Mobile, and Verizon generally retained 30%-40% of each vendor charge placed.
- Despite industry assertions that fraudulent third-party wireless billing was a "de minimis" problem, wireless cramming has been widespread and has likely cost consumers hundreds of millions of dollars.
- The wireless industry was on notice at least as early as 2008 about significant wireless cramming concerns and problems with third-party vendor marketing tactics, yet carriers' anti-cramming policies and sometimes lax oversight left wide gaps in consumer protection:
    - Consumer billing authorization requirements known as the "double opt-in" that were touted as safeguards by industry were porous, and multitudes of scammers appeared to have repeatedly skirted them.
    - Some carrier policies allowed vendors to continue billing consumers even when the vendors had several months of consecutively high consumer refund rates– and documents obtained by the Committee indicate this practice occurred despite

vendor refund rates that at times topped 50% of monthly revenues.

- Carriers placed questionable reliance on billing aggregators in monitoring conduct of vendors that were charging consumers on carriers' billing platforms.

In November 2013, the Attorney General of Texas brought an action alleging that Mobile Messenger, one of the major PSMS billing aggregators, had engaged in a deceptive scheme with vendors to cram consumers' bills. Within weeks – and after years of wireless industry attestations about its effective consumer protection practices – AT&T, Sprint, T-Mobile, and Verizon abruptly announced they would virtually eliminate PSMS billing on their platforms.

Today, while the major carriers have phased out commercial PSMS services, they continue to allow third-party charges on consumers' wireless bills using methods that do not involve PSMS. These include methods sometimes labeled "direct carrier billing" (DCB) through which vendors using websites and apps connect to carrier billing platforms. To date, products and services charged through these non-PSMS billing methods have primarily involved digital content, such as music and apps including games with in-app purchasing capabilities.

Direct carrier billing methods are relatively nascent, and it is not possible at this stage to predict the extent to which scammers will find ways to cram charges on wireless bills under these non- PSMS systems. As new third-party wireless billing methods continue to evolve, it is important that industry and policymakers evaluate the consumer protection gaps that have enabled widespread deceptive and fraudulent charges to be placed on consumers' landline and wireless bills, and to ensure that the unfortunate history of cramming on consumer phone bills does not repeat yet again.

I. Background

A. Initiation of Third-Party Billing on Telephone Bills

Third-party billing on consumer phone bills grew out of two regulatory steps that occurred in the 1980s: the divestiture of AT&T in 1984 and de-tariffing of telephone billing and collection in 1986. Prior to those steps, AT&T had its own billing and collection system that encompassed both local and long-distance charges. Following the break-up of AT&T, regional bell operating companies, also known as local exchange carriers, were not allowed to offer their own long-distance services, and began providing billing collection services to AT&T and other companies that offered long-distance services.[1]

Over time, telephone companies opened these billing platforms to an array of other third- party vendors that offered products and services beyond those directly related to phone service – from webhosting, to online gaming, online photo storage, and roadside assistance.[2] Telephone numbers thus became a payment method similar to credit card numbers. However, third-party charges levied on the phone bill platform did not receive the same protections as credit card payments. For example, with credit card payments, consumers' liability for unauthorized charges is limited to $50, consumers have the right to dispute unauthorized charges, and consumers have the right to seek to reverse a charge.[3] Further, unlike credit card numbers, telephone numbers for landline phones are widely accessible to anyone with a telephone directory.[4]

B. Third-Party Charges on Landline Phone Bills

From early on, industry representatives pledged that voluntary industry practices would protect consumers from billing scams relating to third-party charges on the carrier billing platforms, and carriers agreed upon a set of nonbinding guidelines.[5] At a Senate hearing in July 1998, the President of the United States Telephone Association asserted, "I have a high degree of confidence that these voluntary guidelines will produce an effective means to curb this abuse," that the industry has "a powerful self-interest to correct this problem," and, that the industry was "working overtime" to eliminate "this scourge."[6]

However, over the decade that followed, consumers increasingly began to complain that the third-party charges appearing on their wireline – also known as "landline" – telephone bills were unauthorized. This came to be known as "cramming." State and federal law enforcement agencies brought dozens of enforcement actions against third-party crammers that highlighted problems consumers were encountering. For example:

- In 2006, the Attorney General of Florida filed a lawsuit against Email Discount Network for charging 20,000 Florida consumers' telephone bills for e-mail accounts and coupons they did not request or use;[7]

- In 2009, the Attorney General of Illinois filed a lawsuit against US Credit Find for placing "unauthorized charges on more than 9,000 Illinois consumers' phone bills" for a purported online tutorial that would "help consumers fix their credit;"[8] and
- In 2010, a federal district court awarded the FTC a $37.9 million judgment against Inc21.com Corporation and related third-party vendors after learning that as few as 3% of the defendants' customer base expressly authorized the defendants' charges on their telephone bills.[9]

In 2010, Chairman Rockefeller opened an investigation to examine the extent of third- party billing on landline telephone bills. This investigation resulted in a majority staff report issued in July 2011 that found:

(1) third-party billing on wireline telephone bills was a billion-dollar industry, with over $10 billion dollars in charges placed on consumer bills over a five year period;
(2) a substantial percentage of the charges placed on consumers' telephone bills were likely unauthorized;
(3) telephone companies profited from cramming, generating over $1 billion dollars in revenue from placing third-party charges on customer bills over preceding years;
(4) cramming affected every segment of the landline telephone customer base, from individuals to small businesses, non-profits, corporations, government agencies, and educational institutions;
(5) many third-party vendors were illegitimate and created solely to exploit third-party billing;
(6) many telephone customers who were crammed did not receive help from their telephone companies; and
(7) telephone companies were aware that cramming was a major problem on their third- party billing systems.[10]

Following release of the investigation's findings at a Committee hearing and through a majority staff report, in early 2012 the three major telephone companies – Verizon, AT&T, and CenturyLink – agreed to stop placing third-party charges for enhanced services on their customers' wireline telephone bills.[11] These and other carriers continued, however, to allow third parties to place charges on consumers' *wireless* telephone bills.

C. The Emergence of Cramming on Wireless Phone Bills

Over the past two decades, consumers have migrated from using landline phones to relying on mobile phones,[12] including Internet-enabled smartphones that today represents over half of the mobile phone market.[13] As use of wireless phones began to increase, reports began to mount that consumers were being "crammed," or charged for text message services for which they had not enrolled, on their wireless phone bills. Many of the products that were the subject of consumer complaints were charges for subscription services such as celebrity gossip, horoscopes, sports scores, love tips, and diet tips, which were similar to many of the services found to be fraudulent in the Committee's 2011 wireline cramming investigation.[14]

In recent years, private parties, state Attorneys General, the Federal Trade Commission (FTC), and the Federal Communications Commission (FCC) have brought a number of actions highlighting consumer protection vulnerabilities in the wireless billing system, particularly with respect to charges placed through a system known as premium short message service (PSMS).

For example, between 2008 and 2010, the Attorney General of Florida reached settlements with AT&T Mobility, Verizon, T-Mobile, and Sprint, wherein the companies agreed to issue refunds to customers billed for ringtones, wallpapers, and other mobile content that had been advertised on the internet as free, but resulted in consumers being signed up for monthly text message subscriptions.[15] A plethora of other actions followed.[16]

Most recently, the FTC filed its first wireless cramming complaint against a major carrier, alleging that T-Mobile placed unauthorized third-party charges on its customers' wireless bills, including in some cases, for services that had refund rates of up to 40% in a month. The complaint alleged that T-Mobile knew or should have known that these charges were not authorized and that T-Mobile's billing practices – allegedly burying charges deep into phone bills and without clear descriptions – made it difficult for consumers to find these unauthorized charges on their bills. According to the complaint, when consumers found these charges on their bills, T-Mobile failed to provide full refunds, and directed consumers to the third-party content providers for redress.[17] The FCC announced that it is also investigating complaints against T- Mobile regarding these same practices.[18]

D. State and Federal Enforcement and Regulatory Authority

Agencies at the state and federal level have enforcement and regulatory authority to protect consumers from cramming. Many states have enacted legislation and regulations prohibiting cramming on landline service.[19] Further, California has adopted regulatory provisions specifically addressing wireless cramming.[20] In addition, state Attorneys General have been active in pursuing cases against carriers, billing aggregators, and third-party content providers alleged to have crammed consumers on their wireless bills under their state laws prohibiting unfair and deceptive trade practices.[21]

At the federal level, both the FTC and the FCC have jurisdiction over cramming. The FTC enforces Section 5(a) of the FTC Act, which prohibits "unfair or deceptive acts or practices in or affecting commerce."[22] The FTC has pursued enforcement actions against third-party content providers, billing aggregators, and carriers based on this authority, finding that cramming charges onto phone bills is both an unfair and deceptive practice.[23]

In addition to these enforcement actions, the FTC has held a workshop regarding wireless cramming and explored the possibility of federal regulations.[24]

The FCC has pursued cramming cases under Section 201(b) of the Communications Act of 1934, which states in pertinent part: "[a]ll charges, practices, classifications, and regulations for and in connection with [interstate or foreign] communication service [by wire or radio], shall be just and reasonable, and any such charge, practice, classification, or regulation that is unjust or unreasonable is declared to be unlawful . . . ."[25] The FCC has found "cramming" to be an "unjust and unreasonable" practice.[26]

Current FCC regulations also contain "truth-in-billing" rules regarding both wireline and wireless phone bills.[27] Further, on April 27, 2012, the FCC issued a Further Notice of Proposed Rulemaking (FNPRM) seeking comments on additional measures to prevent wireline cramming and on possible regulatory and non-regulatory measures to address wireless cramming.[28] The comment period closed in July 2012.[29]

In joint comments made to the FCC in 2012, consumer advocates including the Consumers' Union, Consumer Federation of America, and National Consumer League,[30] pressed the agency to adopt rules that would, among other things: (1) prohibit third party charges on wireless accounts except for charitable or political giving;[31] (2) for recurring charges (such as subscriptions), require authorization every time a charge is placed on the consumer's account;[32] (3) require carriers to report wireless cramming complaints on a regular basis;[33] and establish a clear dispute resolution process when consumers complain of unauthorized charges on their wireless bills that includes consumer protections such as the right to withhold payment for the charge while the dispute resolution process takes place.[34] Industry representatives, on the other hand, submitted comments arguing that, at the time, wireless cramming was not "a prevalent consumer issue," that voluntary industry measures would ensure that it did not become a significant consumer issue, and that the FCC lacked authority to issue wireless cramming rules.[35]

On August 27, 2013, the FCC released a Public Notice seeking to refresh the record on cramming "in light of developments and additional evidence"[36] related to both wireline and wireless cramming. The FCC rulemaking remains pending.

II. COMMITTEE INVESTIGATION

In June 2012, Chairman Rockefeller followed up on his wireline cramming investigation to open an inquiry into the scope of unauthorized third-party charges in the wireless context and what steps carriers had undertaken to protect consumers from cramming. He launched this investigation with letters to the four major U.S. wireless phone companies Sprint, T-Mobile, Verizon Wireless, and AT&T[37] requesting information regarding the companies' relationships with third-party vendors and billing aggregators and their practices to prevent cramming on consumer wireless bills.

As evidence of wireless cramming continued to mount, Chairman Rockefeller followed up with additional letters to the same four carriers in March 2013 requesting billing data the companies had provided to the California Public Utilities Commission (CPUC) under California's law requiring disclosures relating to wireless billing.[38] Also in March 2013, the Chairman requested information from five major billing aggregators – Ericsson, mBlox, Mobile Messenger, Motricity, and OpenMarket – relating to their practices in facilitating third-party wireless billing and steps they were taking to prevent abuses.[39]

In June 2013, Chairman Rockefeller wrote the four major carriers to request additional information on questions that had emerged regarding how carriers were monitoring consumer authorizations of third-party billing and following up on consumer concerns.[40]

In November 2013, the Attorney General of Texas filed a complaint against Mobile Messenger, one of the major wireless billing aggregators, alleging that the company had engaged in a deceptive scheme with third-party vendors to cram consumers.[41] These allegations raised questions regarding representations Mobile Messenger had made to the Committee about the company's commitment to consumer protection and the assurances major carriers had given the Committee that aggregators worked with carriers to promote consumer protections in the third- party wireless billing process. In late November, Chairman Rockefeller wrote to Mobile Messenger seeking additional information concerning a subset of vendors whose conduct had raised concerns and pressing for production of previously requested information.[42]

When Mobile Messenger refused to provide key information requested in the Chairman's March 2013 and November 2013 letters, the Committee on March 14, 2014, issued a subpoena to the company, and Mobile Messenger was responsive to the subpoena.

Over the course of the Committee's investigation, Committee majority staff reviewed thousands of pages of narrative and documentary materials produced by wireless carriers and billing aggregators, and conducted interviews of carrier and aggregator representatives as well as other experts. An association for the wireless industry, CTIA-The Wireless Association (CTIA), also provided the Committee documentary and narrative information about the third-party wireless billing system.

III. Overview of Premium Short Message Service (PSMS) Wireless Billing

From the early days of third-party wireless billing, major carriers allowed third-party vendors to charge for their goods and services on customers' wireless accounts. One system that became prevalent is known as the premium short message service (PSMS) whereby consumers would be charged at a higher rate for one-time content or subscriptions received via text message as compared to the standard messaging rate.[43] PSMS charges, along with other third-party charges, are billed to the consumers' wireless account and appear on their billing statement. Over the past few years, use of PSMS has been waning and major carriers ultimately stopped most commercial PSMS billing early in 2014.[44] At the same time, use of other methods that do not involve

PSMS for placing third-party charges on consumers' wireless bills has been increasing.

This section of the report provides an overview of the billing process associated with PSMS and the self-regulation regime that the wireless industry developed to oversee marketing and billing under the PSMS system. Section V of the report addresses alternative third-party billing methods that have been emerging amid the recent decrease in PSMS billing.

A. The PSMS Third-Party Wireless Billing Process

Third-party PSMS billing generally has involved three types of companies: vendors (often known as content providers), wireless carriers, and middlemen known as "billing aggregators" who have provided technology to link content providers and wireless carriers. Under this system, vendors contract with billing aggregators to facilitate placement of charges for goods and services – often referred to as "programs" – on consumers' wireless accounts. Billing aggregators in turn contract directly with the wireless carriers, which control access to the consumers' wireless bills. Each party in this process has retained a portion of the charges paid by consumers.[45]

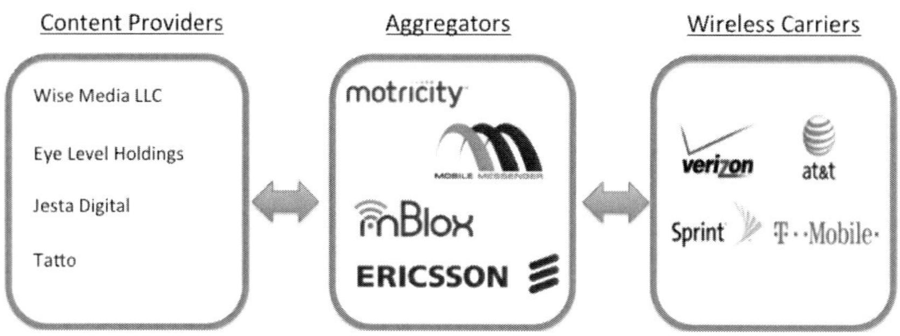

Figure I. Parties in the PSMS Billing Process.

In order for a content provider to send commercial premium text messages, the provider first has to obtain authorization to use a five or six-digit code known as a "common shortcode" (CSC).[46] CTIA-The Wireless Association has managed and controlled issuance of CSCs.[47] Once a content provider has been granted a shortcode, they also must apply to individual wireless carriers to obtain access to the carrier's billing platform to charge consumers for specific content – or "campaigns" – associated with the shortcode.[48]

From a consumer's perspective, the PSMS purchase process as prescribed by industry guidelines has worked as follows. Using an authorization process known as the "double opt-in," consumers must take two affirmative acts when purchasing goods or services with their mobile phone: one to initiate the purchase and one to confirm the purchase.[49] At least one of these actions must be performed using the mobile device associated with the wireless account to be charged.

Industry guidelines also have required content providers to provide information and disclosures to consumers before completing the PSMS charge including the identity of the content provider, contact details for the content provider, a short description of the program, pricing, and terms under which consumers could opt out of the subscription, among other requirements.[50]

Figure II. Steps in Prescribed PSMS Billing Process.[52]

In addition, content providers must provide a confirmation message after affirmative consumer acceptance, including disclosures about the premium charge billed or deducted from the user's account.[51]

Following is an example of what the prescribed authorization process looks like from a consumer's perspective: a consumer would see an advertisement online, on television, or in- store, for downloading a song. The advertisement denotes the advertisement's sponsor, a description of the service or good being offered, its cost, the frequency of the service – which in this case was one song – information regarding customer support, opt-out information, and information regarding any additional carrier costs.

The advertisement would tell the consumer to send a text to the shortcode "12345" with the message "music" to buy the song list in the ad. The consumer would take this step, then receive a message confirming the content ordered, which would reiterate much of the information provided in the original advertisement, including program sponsor, price, frequency of product, how to ask for help with the product purchase, and any additional carrier costs. After confirming this content was accurate, the consumer was to authorize the purchase by sending an affirmative message, in this case "Yes," to the "12345" shortcode. The consumer would then receive a link to the product purchased.

B. Voluntary Industry Oversight over Third-Party Wireless Billing Practices

With respect to third-party billing via PSMS, the U.S. wireless industry developed industry-wide consumer protection standards. Industry-based member organizations created guidelines and recommendations for mobile marketers including parties involved in the marketing and sale of products consumers charge to the wireless phone bills through the PSMS system. Further, carriers developed their own individual policies for oversight of these charges. The following is a description of industry policies concerning the placement of third-party charges on consumer wireless bills.

1. Industry-Wide Oversight

The Mobile Marketing Association (MMA) and CTIA – The Wireless Association (CTIA) spearheaded a number of industry initiatives that were widely adopted throughout the industry for PSMS billing.[53] MMA drafted the Global Code of Conduct and the U.S. Consumer Best Practices for Messaging to provide advertisers, aggregators, application providers, carriers, content providers, and publishers with guidelines for implementing shortcode programs.[54] The guidelines provide detailed requirements for advertising and notice to consumers, along with the appropriate methods for authenticating consumer PSMS purchases.

CTIA – in its Mobile Commerce Compliance Handbook – provides "a unified standard of compliance for mobile carrier billing." The guidelines set forth principles for acceptable program content, opt-in procedures, and cancellation. Many of these are highlighted in the "Consumer Bill of Rights," which provide:

- Programs must use a two-factor authentication for all opt-ins.
- After opt-in, users should receive purchase confirmation of their purchase, either on an additional screen or via a text message.
- All offers must display clear, legible pricing information adjacent to the call-to- action. Pricing information must appear on all screens in the purchase flow.
- Billing frequency information should appear with pricing information, and subscriptions should be labeled clearly as such.
- Clear opt-out instructions must be provided before the purchase is completed and before renewal billing each month.
- All offers must include customer care contact information in the form of a toll-free phone number or an email address. Contact information should function and result in actual user help.
- All offers must supply privacy policy access.
- Purchase flows should include clear descriptions of products offered, and products marketed must match products delivered.
- Product descriptions on customers' wireless bills must reflect accurately the product purchased. Descriptions should include the billing shortcode and the program name.[55]

CTIA in conjunction with an outside auditor would vet content providers that were seeking to lease shortcodes to market and charge products to consumers.[56] Content providers that passed CTIA screening through the Common Short Code Administration could lease a shortcode from CTIA consistent with terms of a user agreement requiring compliance with industry best practices and standards, such as whether the vendor makes clear disclosures to the consumer about how to authorize purchases, or whether the consumer is signing up for a one- shot versus a recurring charge.[57]

Content providers that are permitted to charge consumers via the PSMS system have been subject to ongoing CTIA monitoring for compliance with industry standards surrounding program content, as well as opt-in and cancellation procedures.[58] In 2010, CTIA began providing carriers and billing aggregators access to an online portal that provided the results of these reviews – or audits – in reports that detailed why and how guidelines were violated and that assigned a severity level to each failure. Under this system, each carrier has been responsible for determining what follow up they would conduct with the violating vendor.[59] In addition, carriers receive email notification of new audit findings[60] and weekly reports aggregating the audit failures across the mobile content market.[61] These weekly reports have been compiled into monthly reports to the carriers, which also identify the PSMS billing aggregators that hosted content with the most failures.[62]

2. Individual Carrier Policies

In responses to Committee inquiries, the four major carriers all reported that they comply with CTIA and MMA guidelines for third-party wireless billing, [63] and contractually require the same from their billing aggregators and vendors.[64] All carriers also highlighted several key components of their oversight policies:

- Vetting of third-party vendors and their services beyond the CTIA vetting process; [65]
- The two-step authentication process known as the "double-opt-in" required for consumer approval of third-party services charged on wireless bills[66] (see discussion above in part III.A);
- Monitoring of third-party vendor opt-in and opt-out functionality as well as how they market to consumers; [67]
- Monitoring of third-party vendors through consumer complaint and refund thresholds;[68] and
- Offering consumers the option to block third-party purchases that, when implemented, restrict the purchase of any third-party content billed to a customers' mobile device. [69]

Three of the four carriers said they also have used "first call" resolution of consumer cramming complaints, in which the consumer is generally refunded their money on the first complaint call.[70]

IV. COMMITTEE FINDINGS ON PSMS THIRD-PARTY WIRELESS CRAMMING

Similar to telecom industry assurances about self-regulation of landline billing, from the outset of the Committee's review of cramming on wireless bills, the four major wireless carriers – AT&T, T-Mobile, Verizon, and Sprint – told the Committee that their procedures and practices effectively insulate consumers from cramming on charges incurred through the PSMS system. In July 2012 letters to the Committee, carriers characterized this voluntary system as a "robust process designed to protect customers from unscrupulous actors,"[71] asserting that it provides consumers "simplicity and security,"[72] that the outcome has been a "consistent, secure, and reliable experience" for the consumer,[73] and that carriers have "every incentive to avoid losing a customer due to unauthorized third-party charges."[74] In July 2013 letters to the Chairman, all four carriers asserted that they had only strengthened their anti-cramming practices.[75]

Over this same time period, major industry associations echoed these assurances.[76] In June 2012 comments to the Federal Communications Commission, CTIA-The Wireless Association said that "the wireless industry is already successfully engaged in voluntary initiatives to prevent cramming," calling unauthorized third-party wireless billing a "de minimis" problem.[77] Similarly, the Mobile Marketing Association asserted in May 2013 that the CTIA and MMA rules "are very effective."[78]

However, documents and other information the Committee obtained and reviewed over the course of its inquiry indicate that – just as with landline cramming – industry has gained substantial profits from third-party wireless billing while providing consumers inadequate protections against deceptive and fraudulent charges on their wireless bills. This section details the findings of the Committee majority staff.

A. Carriers Have Profited Tremendously from Third-Party Wireless Billing

It has been estimated that third-party wireless billing activities likely constitute a multi- billion dollar industry.[79] The evidence reviewed by the Committee staff for a sample time frame between 2011 and 2013 supports that analysis.

For example, one carrier reported that nearly $250 million worth of PSMS charges were charged to its customers' accounts in 2011 alone, while another reported over $375 million in total charges for the same year.[80] In addition, information provided by billing aggregators to the Committee shows that the combined revenues of content providers that had relationships with four top aggregators over 2011-2013 totaled over $1.2 billion.[81] This amount – while substantial – does not reflect the entirety of the third-party wireless billing market, as multiple other aggregators were operating in the PSMS market during this time period,[82] and other non-PSMS third-party billing mechanisms were emerging as well.[83]

Further, information provided by the California Public Utility Commission (CPUC) shows that in 2012, over $191 million worth of third-party charges were placed on consumers' wireless bills in California alone – and California has been estimated to constitute about 10% of the wireless market in the United States. Extrapolating and applying the California data across all 50 states, over a span of years, it is likely these numbers would climb into the billions.[84]

Information provided to the Committee by individual carriers indicates that major carriers reaped hundreds of millions of dollars annually from their role in placing third-party charges on wireless phone bills. Contracts reviewed by the Committee show that the carriers generally collected 30% to 40% of the total value of the charges placed.[85] Individual charges are generally small – most often ranging from $1 to $20, with frequent reports of a $9.99 recurring monthly charge. However, the high volume of these charges yields substantial cumulative revenues. For example, one carrier reported processing over 120 million individual third-party transactions on consumer wireless bills in 2011.

In addition to the carriers' revenue shares, contracts reviewed by Committee staff show that certain carriers have collected additional fees that could also add to their profits. For example, one carrier also collected "Excessive Premium Campaign Refund Rate Fees." These additional fees allow the carrier to charge $10.00 per customer care call once a content provider's refund rate exceeded 15% per month.[86] Another carrier has imposed fees ranging from $25,000 to $100,000 where providers experience billing issues which include high levels of refunds.[87]

B. Wireless Cramming has Likely Cost Consumers Hundreds of Millions of Dollars

The evidence reviewed by Committee staff indicates that wireless cramming has likely cost consumers hundreds of millions dollars over the past several years. This assessment is based on a review of data regarding refund rates, consumer complaint information provided by carriers and billing aggregators regarding unauthorized third-party charges, and a number of studies and law enforcement actions that have quantified the extent of wireless cramming.

1. Refund Rates

Beginning in 2011, the California Public Utilities Commission (CPUC) required wireless carriers to provide data about refunds made directly to consumers. Numbers provided by CPUC show that between 2011 and 2013, carriers returned over $60 million in refunds to customers out of $495 million in total third-party wireless charges, just with respect to California wireless consumers.[88] While industry argues that refund rates are a "flawed metric" because refunds can be made for reasons other than cramming,[89] CPUC explained its rationale for using this measure as follows:

> [W]e use refunds as a proxy for complaints because when we had complaint reporting, we would end up in endless semantic digressions around the meaning of the word complaint. So refund is something a little more tangible and we assume that in most cases refunds are not made out of the blue but in relation to some expression of dissatisfaction by the customer.[90]

CPUC also notes that this approach addresses concerns carriers have expressed that "tallying subscriber complaints of unauthorized charges would be excessively burdensome."[91]

As noted earlier, the CPUC numbers concern activity on wireless accounts solely in California, which reflects approximately 10% of the total U.S. wireless market.[92] If the rate of refunds and total charges billed reported to the CPUC were applied nationwide, the total refunds would likely have been well over $200 million out of $1.9 billion in 2012 alone. Even assuming that a portion of the refunds reported to the CPUC are not related to cramming, these numbers provide substantial evidence that cramming on wireless bills has been a serious problem.

Industry argues that numbers of refunds is not an accurate tool to assess the incidence of cramming due to the carriers very liberal refund policies. However, one carrier was able to provide a rough breakdown of refunds specifically attributable to complaints that charges were unauthorized – a category they titled "Authorization of Charge Disputed" – and the results are still high. This carrier reported that 28.1% of total refunds issued in 2012 constituted the "Authorization of Charge Disputed" category.[93] If this percentage were applied to the $200 million figure estimated above to reflect total nationwide wireless refunds for 2012, refunds attributable to cramming for that year would top $50 million nationwide in one year alone. Based on this analysis, over time, wireless cramming has likely cost American consumers hundreds of millions of dollars.

2. Consumer Complaint Data

Consumers also reported on wireless cramming via complaints to federal and state law enforcement as well as to carriers and billing aggregators. One billing aggregator reported receiving over 7,000 contacts from consumers in 2012 alone, over 30% of which involved requests for refunds.[94] A FTC review of complaints from the Consumer Sentinel database, one of the major national resources for compiling local, state, and federal consumer complaints, showed over 2,000 complaints of unauthorized charges on wireless bills between 2010 and 2013.[95]

In addition, in a 2013 survey conducted by the Office of the Attorney General of Vermont, 60% of respondents reported that the third-party charges found on their wireless telephone bills were unauthorized.[96]

Industry representatives have argued that complaint numbers were low and that the incidence of cramming on wireless bills was insignificant.[97] However, consumer complaint tallies likely reflect numbers far lower than actual cramming occurrences. Evidence shows that consumers are frequently unaware that third-party charges are appearing on their telephone bills. FTC Commissioner Maureen Ohlhausen elaborated on this point as follows:

> Indeed, we are aware of thousands of consumer complaints about unauthorized charges on wireless bills. And we believe that these complaints may well under represent the problem or under report the problem. From surveys done in the landline cramming context, we know that many consumers are unaware that third parties can place charges on their phone bills. We also know that consumers often fail to spot unauthorized charges on their bills. They may simply look at the overall bill amount and pay in full without doing a line-by-line review; or they may read the bill and fail to spot the charges because they're buried deeply within the bill or listed in generic sounding categories, such as premium services.[98]

Committee staff review of consumer complaints substantiates this viewpoint. Individual consumers often reported only finding the charges after paying them for extensive periods of time. For example, one consumer complained, "I was billed for 18 months for $9.99 ($10.76 after taxes) for something I had no clue about and up till [sic] today when I reviewed my bill I noticed these charges." Another consumer reported, "Having automatically paid my bills for one year, I unfortunately just learned I was paying for unsolicited text messages for over a year."[99]

Indeed, the complaint filed by the FTC against T-Mobile alleges that the bill statements received by customers did not adequately disclose PSMS charges.[100]

Customers reviewing their bill online allegedly could not see these charges by viewing a summary of the charges; only by clicking a series of links could they find premium service charges.[101] Figure III below is a graphic from the FTC complaint in this case illustrating how the charges allegedly were shown on the consumers' paper statements.

Premium charges were not individually listed in the summary section of the bill. Though they were itemized in a "Premium Services" section several pages into the bill, the information was presented in a way that did not adequately explain that the charge was for a recurring subscription service authorized by the consumer.[103]

If these allegations are true, it is entirely possible that many consumers over a number of years had paid for third-party charges they did not authorize.

3. State and Federal Actions

The charges detailed in numerous state and federal law enforcement actions also underscore the broad consumer impact of wireless cramming. These cases have charged that consumers have been victims of cramming schemes costing them hundreds of millions of dollars. For example:

- In 2011, the Attorney General of Texas filed a lawsuit against Eye Level Holdings, LLC, alleging that the defendants collected millions of dollars through the placement of unauthorized charges on the wireless telephone bills of thousands of Texas residents.[104]
- In 2013, Wise Media and its owners agreed to settle FTC allegations that they caused more than $10 million in consumer harm by placing unauthorized recurring $9.99 monthly fees on consumers' wireless bills.[105]
- In June 2014, a district court issued a stipulated order for a monetary judgment totaling over $150 million in a case brought by FTC alleging defendants used deceptive websites to cram consumer's wireless bills.[106]

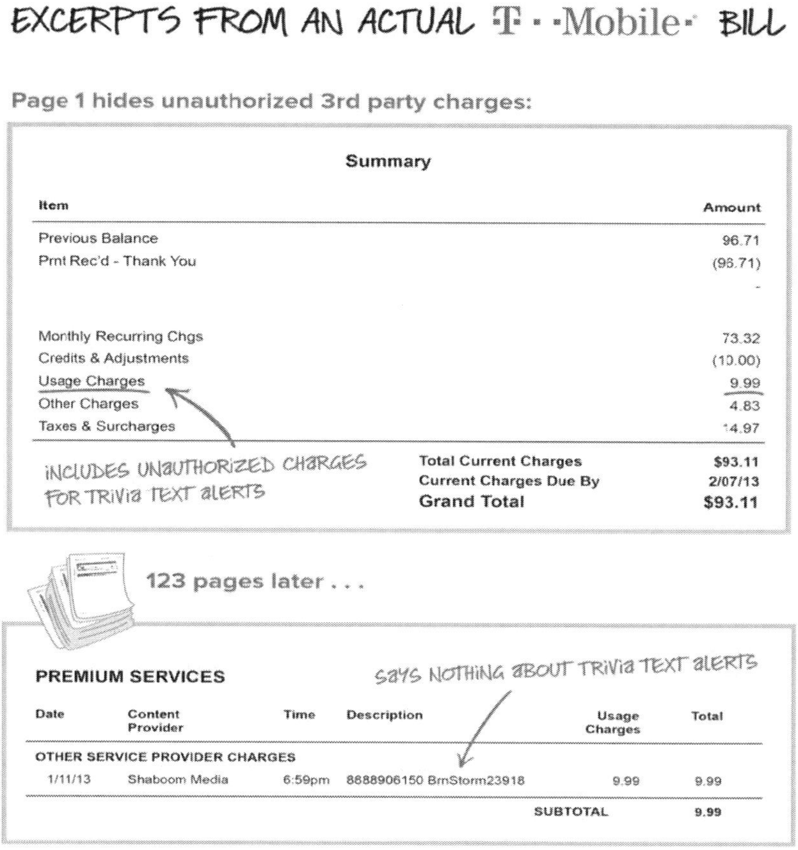

Figure III. T-Mobile Bill Summary.[102]

- In July 2014, the FTC charged T-Mobile with placing third-party charges on consumers' wireless bills despite clear warnings that the charges were unauthorized, and engaging in billing practices that made it difficult for consumers' to discern fraudulent charges, alleging that these practices cost consumer millions of dollars in injury.[107]

Indeed, review of 2011-2013 data provided to the Committee by major billing aggregators regarding revenue generated by content provider clients shows that many of the top revenue generators in this time frame were ultimately the subject of state or federal enforcement actions. According to this data, the subjects of enforcement actions generated approximately 23.5% of total revenue reported to the Committee for this time period – $289,037,831 of $1.2 billion.[108]

In short, the cumulative evidence revealed by enforcement actions, consumer complaints, refund rates, and studies, indicates that hundreds of millions of dollars in crammed charges have been placed on consumers' wireless bills over the past several years.

C. Carriers Were on Notice about Cramming and Other Vendor Problems

Carriers should have known at least as early as 2008 that consumers were complaining of cramming on their wireless bills. Beginning in 2008, the Florida Attorney General entered into enforcement settlements with Cingular (AT&T), Verizon, Sprint, and T-Mobile over allegations that unauthorized charges had been placed on their consumers' bills.[109] These settlements created a "best practices" regime intended to ensure that consumers were receiving clear and conspicuous prices and terms of the content being purchased before such charges could be placed on a consumer's wireless bill.[110]

According to CTIA, shortly after the last settlement was signed in October 2010,[111] this best practice regime was incorporated into the mobile billing standards against which the industry audited vendors for compliance.[112] The CTIA audit reports that followed indicated that, three years after the Florida enforcement cases sounded the alarm about wireless cramming, the overwhelming majority of vendors allowed to charge consumers on wireless billing platforms were not meeting basic standards.

For example, with respect to the monthly reports CTIA provided carriers summarizing in- market auditing, the January 2011 report showed a failure rate of nearly 100% for marketing offers tested – or "intercepted" – by the auditors.[113] Virtually all of the failures[114] were for violations classified by CTIA as "severity level one," meaning "serious consumer harm."[115] While monthly industry audit reports after January 2011 showed declining numbers of compliance failures, it was not until September 2011 that more interceptions were reported to have passed than failed.[116] And in January 2012, audits still showed a 25% failure rate.[117] After August 2012, the reports indicated passage rates of 95% or higher, meaning that 95% of offers tested complied with CTIA guidelines.[118]

D. Industry Self-Regulation Has Left Gaps in Consumer Protection

1. The Double Opt-In Safeguard Was Porous

Many of the voluntary policies and practices industry instituted to protect against cramming in the PSMS system are similar to those industry touted in the landline context.[119] However, as with law enforcement actions in the 2000s involving landline cramming,[120] a series of recent state and federal law enforcement cases concerning wireless cramming have highlighted potential vulnerabilities with industry's voluntary consumer protection system. In particular, recent actions raise concerns regarding the effectiveness of the double opt-in authorization.

As discussed above, industry representatives have argued that a key protection against wireless cramming that was not present in the landline context is the "double opt-in" requirement,[121] as it involves affirmative steps by the consumer that are "immediate," "current," and "actionable" before billing can be activated.[122] However, several cases brought at the state and federal level in the last few years have detailed multiple ways content providers have circumvented the double opt-in. For example:

- According to an FTC action brought in December 2013, content providers operated a scam in which they billed consumers for services that were not authorized through the use of misleading websites. The complaint cites as an example a website that offered to sign up consumers for Justin Bieber concert tickets if consumers provided their phone number, and alleges defendants likely used that phone information to sign up the consumer for services without their knowledge.[123]
- According to the complaint in a separate FTC action brought in April 2013, consumers received unsolicited text messages from a third-party vendor and were charged on their wireless bills for the vendors' services regardless of whether the consumers had ignored the text message or had responded via text message that they did not want the services.[124]
- A complaint brought by the Attorney General of Texas in 2011 claimed that the defendants used deceptive websites to entice consumers to enter their wireless telephone numbers. According to the complaint, defendants' websites would come up as prominent sponsored links when consumers entered generic search queries for

> information on topics such as "song lyrics." Defendants' link would not mention subscriptions or costs, and if consumers clicked on the link they would be taken to a page where they were encouraged to enter their phone number with prominent instructions such as "enter your cell phone number to access the lyrics" without any clear and conspicuous disclosures that consumers were in fact signing up for paid subscription services. The complaint further alleged that, to conceal this flawed enrollment process from regulators, carriers, and consumers re-visiting the site, defendants created "dummy" websites that included larger, brighter, and clearer disclosures on the service cost and subscription nature.[125]

In another case brought by the Attorney General of Texas, defendants allegedly worked around the "double opt-in" requirement through the following process depicted with the accompanying graphics reproduced in Figure IV below.[126] An online search for Olive Garden coupons would turn up a link for a 50% discount coupon, without disclosures regarding any fees or subscriptions charged for enrolling.[127]

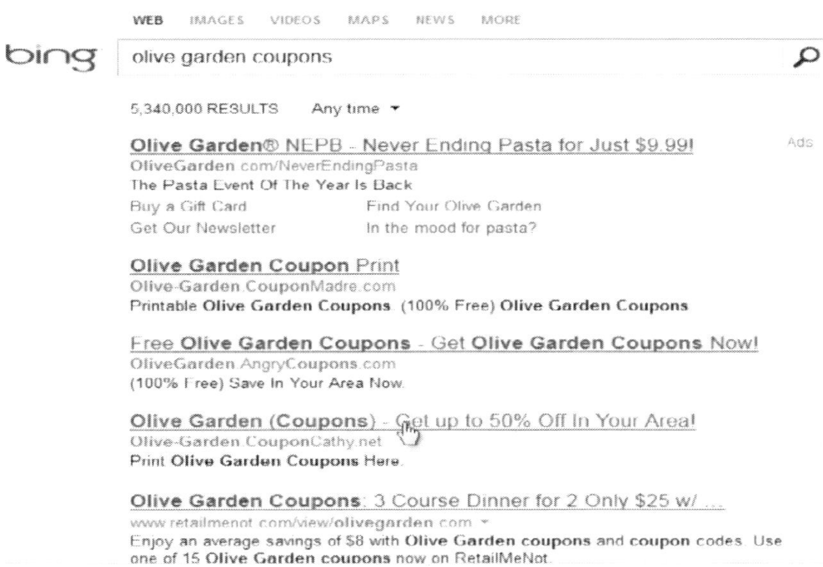

Figure IV. Internet Search Producing PSMS Website.

Consumers who tried to download the coupon were required to enter their personal information including mobile phone number (see depiction of this screen in Figure V below). The complaint alleged that, by entering their mobile phone numbers, consumers unknowingly authorized a $9.99 per month subscription service providing monthly horoscopes.[128] Consumers were not provided clear disclosures regarding the actual offer of the subscription service or its relevant terms and conditions unless consumers scrolled down.[129]

In this case, the Texas Attorney General also alleged that content providers lured unknowing consumers to subscribe for deceptive PSMS campaigns through the use of website addresses that contained common typos and misspellings of the addresses of legitimate websites. These websites would encourage consumers to share personal information including their phone numbers in exchange for a promised gift card.[130]

Industry representatives have underscored that wireless cramming enforcement cases have involved conduct that circumvents consent mechanisms, and that generally the double-opt in mechanism was sound.[131] However, conduct described in the above cases allegedly continued for time periods as long as several years, indicating substantial weaknesses in the wireless industry's ability to root out abuses of consumer authorization requirements.

Figure V. Website Drawing Consumers into PSMS Campaign.

2. Tolerance for High Consumer Refund Rates Raises Questions about Carrier Commitment to Preventing and Addressing Cramming

All four major carriers cited consumer refund thresholds as a tool for spotting potential vendor misconduct. However, the thresholds and response actions triggered by breach of these thresholds varied widely in the policies carriers described to the Committee.[132] Documents produced to the Committee by billing aggregator Mobile Messenger regarding a subset of its vendors provided a further window into the role that refund threshold policies played in the industry's oversight of the PSMS billing system.[133] Review of these documents revealed that carriers saw extremely high refund rates and high monthly refund totals for some vendors and were not consistent in how they followed up on red flags concerning vendor misconduct.

a. Carrier Policies on Refund Thresholds

Verizon Wireless, AT&T, Sprint, and T-Mobile established different refund threshold levels for triggering additional vendor review, and their policies also varied regarding specific prescribed follow-up steps. For example, Verizon Wireless's policy provided that if the refund rate *for any one program* in any month is between 5% and 7.99%, *all* PSMS campaigns managed by that vendor would be suspended, and if the refund rate exceeded 8%, *all* PSMS campaigns of the vendor would be terminated.[134] Billing aggregator documents indicate that the policy applied to shortcodes on Verizon's network was that suspension for refunds between 5% and 7.99% meant a bar on acquiring *new* subscribers for a period of 90 days.[135] The policy applicable once refunds exceeded 8% involved both a bar on new subscribers and a requirement that existing subscribers be unsubscribed for shortcodes with subscriptions that brought in an average revenue of at least $5000 over the previous three months.[136]

AT&T, on the other hand, stated that it did not have a static threshold for refund rates but rather it adjusted the threshold "over time to account for changes in the overall refund rate as observed."[137] The company further stated it had a general disciplinary policy that could involve "suspending or de-provisioning the short code, and/or terminating the content provider" from the carrier's network, but these steps were not tied to specific threshold violations.[138] Billing aggregator documents indicate that as of May 2013, the policy applied to shortcodes on AT&T's network was to "enforce a 30-day suspension on any shortcode with a combination of a failed audit and a refund rate of 18%."[139]

Sprint reported that its policy provided for a "combination of metrics" including refund rates to assess noncompliance, and was penalizing with "lower revenue share" those aggregators that work with vendors demonstrating noncompliance or a refund rate greater than 10%.[140] According to billing aggregator documents from May 2013, the policy applied to shortcodes on Sprint's network was that a refund rate between 0% and 7% meant incentives and bonuses might apply; refunds between 7.01% and 12% merited a "normal payout;" and refunds greater than 12.01% meant Sprint would apply a "25% penalty on the average monthly retail revenue ... for the three-month period and risk of code termination."[141]

Finally, T-Mobile stated that its refund threshold was 15%, at which point aggregator partners and vendors would be "penalized financially in accordance with the terms of the aggregator's contract."[142] Additional detail provided by billing aggregator documents indicates that the policy applied to shortcodes on T-Mobile's network was that T-Mobile would charge a vendor $10 for each refund/customer care call after refund rate surpassed 15%;[143] in addition, T-Mobile would apply a multi-step "Refund Performance Improvement Plan" (PIP) if a vendor's refund rate exceeded 15% and involved at least $10,000 in "excessive refund fees."

The documents indicate that the PIP program involved placing the vendor on a "watch list" for 12 months for remediation steps before T-Mobile would terminate the campaign.[144] Once on the watch list, the vendor had three months to address the high refund rate or else in month four, new subscribers would be suspended for a one-month period. The vendor could resume new subscriptions in month five after this suspension, and had three additional months to address the refund rate. If, after month seven, the vendor still qualified for the "watch list," the PIP program applied a two-month suspension of new subscribers in months eight and nine. The vendor could resume new subscriptions in month 10, but the campaign at issue would be terminated after month 12 if the vendor still met PIP criteria.[145]

It is worth noting that even the lowest stated refund threshold rates of 5%-7.99% that were set forth in policies carriers described to the Committee are substantially higher than the threshold used for chargebacks levels on consumer credit card bills as a trigger for follow-up with the merchant whose chargebacks are at issue. For example, under VISA's chargeback policy, merchants will receive a notification and request for explanation if the ratio of transactions charged back to total transactions exceeds 1%, where the merchant has had over 100 transactions and 100 chargebacks in that month.[146]

Documents received by the Committee from Mobile Messenger included a number of examples demonstrating the follow up actions taken by carriers after adverse audit findings or other red flags regarding particular shortcodes.[147] However, documents also indicated there were instances where carriers were lax in overseeing or enforcing their own stated policies. This issue is illustrated in an email chain between AT&T and Mobile Messenger in October 2013 when AT&T sent Mobile Messenger a notice of termination for content provider Anacapa, a client of Mobile Messenger, due to "excessive CTIA Sev 1 [severity 1] audit failures."[148]

In the course of the email chain, an AT&T representative gives more explanation for the termination. He begins by noting that when Anacapa requested access to AT&T's billing platform in November 2012, Anacapa "did not pass our internal vetting process, … and we rejected them from running PSMS campaigns."[149] However, AT&T nonetheless let one of the Anacapa shortcode campaigns have access to the AT&T billing platform, as in fact AT&T had "failed to reject" that shortcode. Anacapa was able to bill consumers on AT&T's platform well into 2013, despite two AT&T suspensions of the Anacapa shortcode in the first part of 2013. Further, in May 2013 AT&T "drafted" a termination notice but again "failed to deliver" it.[150] The October 2013 email summary also noted that AT&T had found that Anacapa had received twenty severity 1 – "serious consumer harm" – audit findings across several of its shortcodes from October 2012 to October 2013, including two Severity 1 findings on the shortcode that was apparently erroneously allowed to use the AT&T network.[151]

b. Some Vendors Had Exceedingly High Refund Rates That at Times Spanned Several Months

Documents reviewed by the Committee regarding a subset of vendors who contracted with billing aggregator Mobile Messenger indicate that some vendors experienced high monthly refund rates that in some cases topped 50% of monthly revenues and amounted to hundreds of thousands of dollars in refunds in a single month. For example, in a July 2011 Mobile Messenger email to vendor representatives, Mobile Messenger employees reported violations of T-Mobile thresholds on 11 different shortcodes for the preceding month, including one shortcode with a 50.5% refund rate and $55,974 in refunds for the month, and others with 43.7%, 38.4%, and 36.1% rates. The refunds for the 11 listed shortcodes totaled over $450,000 that month.[152]

A similar Mobile Messenger email notification to vendor representatives in October 2012 notes that 11 shortcodes had exceeded AT&T's 18% refund threshold in the preceding month, including one shortcode with a refund ratio of 56.8% and $124,759 in refunds for the month, another with a ratio of 31.4% and $100,949 in refunds. The 11 shortcode refunds that month together totaled nearly $600,000.[153] Documents indicate that other carriers also had high refund rates and high refund totals.[154]

Documents also show that refund rates on the same shortcode at times exceeded carrier thresholds for a number of months at a time. For example, Mobile Messenger sent emails to vendor representatives notifying them that refunds on shortcode 67145 exceeded AT&T's threshold in February 2012 (with a 33.9% refund rate);[155] March 2012 (40.6%);[156] and May 2012 (18.1%).[157] Mobile Messenger sent similar notification emails that refunds on shortcode 85820 exceeded T-Mobile's threshold in December 2010 (20.06%);[158] February 2011 (34.13%);[159] and March 2011 (31.13%).[160]

c. Case Study on Vendor with High Refund Rates: Variation in Carrier Response Underscores Broad Latitude Afforded by the Self-Regulatory System

Documents indicate that different carriers employed different practices regarding follow- up on red flags such as high refund rates and adverse audit findings associated with shortcodes. For example, in October 2011, Verizon wrote to Mobile Messenger regarding several shortcodes used by Mobile Messenger client NeoImage.[161] This group of shortcodes also included certain shortcodes that appeared on high refund rate notices sent by Mobile Messenger to vendor employees concerning all four major carriers in 2011.[162]

Verizon's October 2011 letter requested that, because of "the repeated and serious nature" of content provider violations of requirements concerning premium messaging campaigns, "all of the premium messaging campaigns managed by the content provider must be terminated," and Verizon Wireless "will not consider reactivation of the shortcodes, or any new campaigns from the content provider."[163] The Verizon letter required the content provider to block all new subscriptions to the code and opt-out enrolled customers on a rolling basis "at the time their subscriptions otherwise would be renewed."[164]

Consistent with this letter, billing statements for Mobile Messenger client NeoImage indicate that Verizon ceased allowing NeoImage shortcodes to bill on its platform starting in late 2011.[165] However, the billing statements also indicate that the three other major carriers and many others continued to allow NeoImage to charge their customers through March 2013, the last date of statements provided to the Committee for NeoImage. In 2012, NeoImage charged a total of over $10 million to consumers' wireless bills across different carrier platforms.[166]

The billing statements produced by Mobile Messenger indicate that a number of carriers also continued to allow NeoImage to charge on their platforms for activity on several of the specific shortcodes that Verizon terminated in October 2011. For example, the statement for January 2012 shows that, with respect to campaigns on shortcode 89147, $107,000 in charges were placed with AT&T, $33,800 with T-Mobile, and $21,700 with Sprint.[167] With respect to the same shortcode, the March 2012 billing statement showed $81,100 in charges placed with AT&T, $20,600 with T-Mobile, and $16,300 with Sprint.[168] Documents also indicate that in January 2012 refunds on this same shortcode exceeded refund thresholds for both T-Mobile and AT&T, with a 15.7% refund ratio for T-Mobile,[169] and a 19.01% refund rate for AT&T.[170]

This example regarding NeoImage shortcodes illustrates that carriers had wide discretion in responding to indicia of vendor problems such as high refund rates.

3. Carriers Placed Questionable Reliance on Billing Aggregators As Oversight Partners

In submissions to the Committee in 2012 and 2013, a number of major carriers emphasized the important and reliable role that billing aggregators played in ensuring that vendors comply with consumer authorization requirements and other industry standards. Sprint noted that in its experience, the company's "reward/penalty system influences aggregators to work only with reputable content providers and to ferret out non-compliant PSMS campaigns."[171] AT&T assured the Committee that "in November 2012 the double opt-in procedures of all then existing Billing Aggregators were reviewed and certified" as compliant with the company's consent management program.[172] And T-Mobile asserted last June that "we are aware of no information that aggregator partners have played any role in cramming or otherwise facilitating improper third-party billing."[173]

Major billing aggregators contacted in the Committee's inquiry also attested to their role in the industry's compliance system. For example, Mobile Messenger, one of the leading aggregators, underscored that it is "committed to consumer protection," with a "dedicated compliance team" to review and test vendor campaigns,[174] and that the company has spent "considerable resources" to ensure that the subscription and billing process and the company's content provider and advertiser clients abide by the "robust" industry guidelines.[175]

However, the allegations in the November 2013 action by the Texas Attorney General[176] raise serious questions about the effectiveness of aggregators as partners to carriers in combatting cramming as well as how closely carriers were scrutinizing aggregator practices. As noted above, in this action, the Texas AG alleged that Mobile Messenger was part of a "deceptive scheme" to trick consumers into signing up for unwanted "services" including ringtones, weekly text messages containing horoscopes and celebrity gossip, and coupons. According to the complaint, Mobile Messenger actively assisted content providers with circumventing consumer protections that carriers implemented, including the double opt-in and thresholds relating to consumer complaints and audit reports.[177]

In addition, Mobile Messenger documents reviewed by Committee staff about a subset of Mobile Messenger vendors underscore that the company was in a position to see red flags about worrisome shortcodes and vendors from both the industry-wide audits as well as from reports of individual carrier refund rates and vendor penalties.[178] And yet, in the case study discussed above, after one of the major carriers in October 2011 cut off all business with a vendor that had raised non-compliance concerns and been the subject of high refund rates across major carriers, Mobile Messenger continued doing business with the same vendor through 2013.[179] Such actions raise questions about whether Mobile Messenger served as a rigorous oversight partner with carriers in weeding out vendors with records of non-compliance with industry standards.

V. Emerging Third-Party Wireless Billing Technologies and Potential Consumer Protection Issues

Though commercial PSMS billing has now virtually ended among the major carriers,[180] it is still possible for consumers to buy digital content online and bill those purchases to their wireless phone accounts. This is generally called the "direct carrier billing" payment option. Direct carrier billing is now offered by a variety of U.S. companies for music, applications, games, movies, and television shows, from retailers such as Sony, Facebook, Skype, and Rhapsody.[181]

Direct carrier billing for social media and gaming purchases increased 30% year-over- year from 2009-2012.[182] This option could become even more widely available in the future, for goods and services outside of digital content. According to CTIA, additional entities "currently using or planning to adopt" third-party billing include "major news organizations, companies offering video streaming, gaming companies, parking services, and even pizza delivery services."[183]

In discussions with Committee majority staff, carriers have differentiated between two methods of direct carrier billing: the "storefront" approach and the "billing aggregator" approach. In the "storefront" approach, consumers are given the option of billing their wireless account when making a purchase from a digital distribution platform that offers applications, music, movies, and games created by any number of vendors.[184] Under this billing model, the carriers rely on the company offering the digital distribution platform to vet the vendors who create the digital content offered in their store.[185] One example of the "storefront" billing method is the Google Play store. Since 2011, consumers have been able to make purchases from the Google Play store and bill them to their wireless account.[186]

Outside of the storefront approach, direct carrier billing is also an option for a number of additional vendors who utilize billing aggregators in order to place the charges on consumers' wireless accounts. In this model, both the carriers and aggregators vet each vendor before the vendor is permitted to bill consumers on their wireless accounts.[187] A handful of these billing aggregators have emerged to act as middlemen between vendors and wireless carriers.[188]

With respect to direct carrier billing, to date there are no industry-wide best practices or central monitoring similar to what was in place for PSMS. Instead, oversight of direct carrier billing occurs at the individual carrier level.[189] Policies described by several major carriers in briefings to Committee majority staff include the following features, among others:

- Clear disclosures by the content provider to the consumer regarding the terms of purchase;
- Clear designation of third-party vendor purchases on consumers' phone bills; and
- Carrier monitoring of refund rates and consumer complaints.

Some carriers also said they place caps on third-party purchases from $25 to $80 per month, and for consumers that have more than one wireless line on their plan these caps apply per line.[190]

As of now, direct carrier billing is primarily an option for digital content and only represents a small fraction of purchases made via computers and smartphones.[191] However, as noted by the CTIA in comments to the FCC, U.S. companies are increasingly offering direct carrier billing for purchases.[192] Direct carrier billing is a more widely utilized form of purchase internationally[193] and with the continued growth in the unbanked and underbanked population in the United States, it is conceivable that direct carrier billing could become a more widely utilized payment option in the future.

Currently, major carriers assert that they are seeing minimal indicia of consumer complaints involving direct carrier billing, including very few consumer complaints and refund rates around 1%-1.5%.[194] However, in light of the extensive evidence of cramming that has occurred to date in both the landline and wireless contexts, and the potential that a growing number of consumers may use this payment option in the future, this staff report recommends that industry and policymakers:

- Vigilantly monitor evolving third-party billing practices to make sure that bad actors do not find ways to penetrate barriers to cramming on these new systems; and
- Evaluate consumer protection gaps that occurred in the landline and PSMS contexts to establish consistent policies going forward that will provide consumers with appropriate transparency in the third-party billing process and a clear avenue of recourse where unauthorized charges occur.

End Notes

[1] Senate Committee on Commerce, Science, and Transportation, *Staff Report on Unauthorized Third- Party Charges on Telephone Bills*, at 22 (July 12, 2011).

[2] Senate Committee on Commerce, Science, and Transportation, *Staff Report on Unauthorized Third- Party Charges on Telephone Bills*, at 22 (July 12, 2011).

[3] *See* Fair Credit Billing Act, 15 U.S.C. §§ 1666-1666j; Consumer Credit Protection Act 15 U.S.C. § 1643; Regulation Z, 12 C.F.R. § 1026.13.

[4] *See* Senate Committee on Commerce, Science, and Transportation, *Staff Report on Unauthorized Third- Party Charges on Telephone Bills*, at 2 (July 12, 2011).

[5] *See* Federal Communications Commission, *Anti-Cramming Best Practices Guidelines* (available at www.fcc.gov/Bureaus/Common_Carrier/Other/cramming/cramming.html) (accessed July 7, 2011).

[6] Permanent Subcommittee on Investigations of the Senate Committee on Governmental Affairs, *Hearing on "Cramming": An Emerging Telephone Billing Fraud*, 105th Cong. (July 23, 1998) (S. Hrg. 105-646).

[7] *See* Settlement Agreement, *Florida, Office of the Attorney General v. Email Discount Network*, Fla. 2d Cir. Ct. (No. 2006 CA 2475) (Feb. 15, 2007).

[8] *See* Press Release, *Madigan Reaches Agreement with US Credit Find to Prevent Phone Cramming*, The Office of the Illinois Attorney General (June 18, 2009).

[9] *See Federal Trade Commission v. Inc21.com Corp.*, 745 F.Supp.2d 975, 982-983 (N.D. Cal. 2010).

[10] Senate Committee on Commerce, Science, and Transportation, *Staff Report on Unauthorized Third- Party Charges on Telephone Bills*, at ii-iv (July 12, 2011).

[11] *See* Senate Committee on Commerce, Science, and Transportation, *Rockefeller Hails Verizon Decision to Shut Down Unwanted 3rd-Party Charges on Telephone Bills* (Mar. 21, 2012); Senate Committee on Commerce, Science, and Transportation, *Another Major Phone Company Agrees to End Third-Party Billing on Consumer Phone Bills* (Mar. 28, 2012); *Chairman Rockefeller Introduces Telephone Bill Anti- Cramming Legislation*, U.S. Federal News (June 14, 2012).

[12] A recently released report by the National Center for Health Statistics showed that two out of five U.S. households, or 41%, had only wireless phones in the second half of 2013 (July-December 2013). Pew Research Center, *CDC: Two of Every Five U.S. Households Have Only Wireless Phones* (July 8, 2014) (online at http://www.pewresearch.org/fact-tank/2014/07/08/two-of-every-five-u-s-households-have-only- wireless-phones/).

[13] As of January 2014, 90% of American adults had a cell phone and 58% had a smartphone. Pew Research Center, *Cell Phone and Smartphone Ownership Demographics* (online at http://www.pewinternet.org/data-trend/mobile/cell-phone-and-smartphone-ownership

[14] *See What's Your Sign? It Could Be a Cram*, New York Times (Mar. 24, 2012) (reporting on a consumer who complained of being billed for horoscope text services not authorized). In the wireline cramming investigation, the Committee found that companies that were charging consumers each month for e-mail accounts that included weekly e-mail messages with "celebrity gossip" and "fashion tips." *See* Senate Committee on Commerce, Science, and Transportation, *Staff Report on Unauthorized Third-Party Charges on Telephone Bills*, at ii-iii (July 12, 2011); *See also* footnote 16 *infra*, detailing legal actions concerning various subscription services.

[15] *See FL AG McCollum in Settlement With Sprint Over 'Free' Ringtones*, Bloomberg (Oct. 8, 2010) (online at http://www.bloomberg.com/apps/news?pid=21070001&sid=aXwc4 FpkupsU); *T-Mobile $600k Settlement with Florida AG Affects All Mobile Content Marketing*, Mobile Marketer (July 22, 2010) (online at http://www.mobilemarketer.com/ cms/news/legal-privacy/6873.html). *See* Part I.D below for discussion of additional state and federal actions.

[16] *See Texas v. Eye Level Holdings, LLC, et al.*, Tex. D. Ct., Travis County (No. 1:11-cv-00178) (Mar. 11, 2011) (where the Texas Attorney General accused the defendants of engaging in deceptive trade practices by running a text messaging scheme that cost consumers in Texas millions in unauthorized wireless charges; and defendants agreed to pay nearly $2 million to settle the charges); *Federal Trade Commission v. Wise Media, LLC, et al.*, N.D. Ga. (No. 1:13cv1234) (Apr. 16, 2013) (where the third-party content provider was charged for placing over $10 million on consumers' wireless bills for unauthorized charges for PSMS messages containing horoscopes, love and flirting tips, and other information); *Federal Trade Commission v. Jesta Digital, LLC, also d/b/a JAMSTER*, D.D.C. (No. 1:13-CV-01272) (Aug. 20, 2013) (where the third-party content providers were charged with cramming unwanted charges on consumers' cell phone bills for ringtones and other mobile content); *Texas v. Mobile Messenger U.S. Inc., et al*, Tex. D. Ct., Travis County (Nov. 6, 2013) (alleging that defendants, who were a billing aggregator, four content providers, and an online advertising placement business, conspired to enroll consumers in PSMS programs

for ringtones, horoscopes, celebrity gossip news, and other coupons without consumer consent); and *Federal Trade Commission v. Tatto, et al.*, C.D. Cal (No. 2:13-cv13-8912-DSF-FFM) (Dec. 5, 2013) (in which FTC alleged that defendants placed millions of dollars on consumers' wireless phone bills for text messages that consumers did not authorize; and defendants ultimately agreed to surrender over $10 million in assets to settle these charges). Private parties also have brought legal actions involving third- party cramming charges. *See Tracie McFerren v. AT&T Mobility LLC*, Sup. Ct. of Ga. (No. 08-cv-151322) (May 30, 2008) (a class action suit alleging that AT&T failed to set up controls to stop unauthorized third-party charges on consumers' wireless bills); *Gray v. Mobile Messenger Americas, Inc.*, S.D. Fl. (No. 0:08-cv-61089-CMA) (July 11, 2008) (a class action suit charging Mobile Messenger, a billing aggregator, with placing unauthorized third-party charges on consumers' wireless bills); *Armer v. OpenMarket, Inc.*, W.D. of Wash. (No. 08-CV-01731-CMP) (Dec. 1, 2008) (a lawsuit against OpenMarket, a billing aggregator, and Sprint concerning alleged unauthorized charges for PSMS messages containing content such as ringtones, sports score reports, weather alerts, and horoscopes); and *Cellco Partnership d/b/a Verizon Wireless v. Jason Hope, Eye Level Holdings, LLC, et al.*, D. Ariz. (No. 2:11-cv-00432-DGC) (Mar. 7, 2011) (in which Verizon charged that the third-party content provider collected unauthorized or deceptive charges on consumers' wireless bills through PSMS messages).

[17] *See Federal Trade Commission v. T-Mobile USA, Inc.*, W.D. Wash. (No. 2:14-cv-00967) (July 1, 2014) (online at ftc.gov/enforcement/cases-proceedings/132-3231/t-mobile-usa-inc).

[18] FCC, *FCC Investigates Cramming Complaints Against T-Mobile* (July 1, 2014) (online at http://www.fcc.gov/document/fcc-investigates-cramming-complaints-against-t-mobile).

[19] *See, e.g.*, Mich. Comp. Laws § 484.2502; Cal. Pub. Util. Code § 2890; 52 Pa. Code § 64.23; Tex. Util. Code § 17.151; Va. Code § 56-479.3. In 2011, Vermont became the first state to enact legislation prohibiting third-party billing on landline telephone bills, with three limited exceptions: "(1) billing for goods or services marketed or sold by entities subject to the jurisdiction of the Vermont Public Service Board; (2) billing for direct-dial or dial-around services initiated from the consumer's telephone; and (3) operator-assisted telephone calls, collect calls, or telephone services provided to facilitate communication to or from correctional center inmates." 9 Vt. Stat. § 2466. Illinois enacted similar legislation in 2012. *See* 815 ILCS 505/2HHH.

[20] The California Public Utilities Commission adopted rules that (1) established that wireless carriers must obtain explicit authorization from consumers before they can be billed for third-party charges; (2) establish that the carriers must refund consumers for unauthorized charges and investigate any complaints of unauthorized charges; and (3) requires wireless carriers to report quarterly the total amount of refunds given to California consumers for unauthorized charges and third party vendors that have been suspended or terminated. *See* Press Release, *CPUC Strengthens Consumer Protections Against Cramming and Fraud on Telephone Bills*, California Public Utilities Commission (Oct. 28, 2010).

[21] *See, e.g.* cases cited at footnote 16, *supra*.

[22] 15 U.S.C. § 45(a). Misrepresentations or deceptive omissions of material fact constitute deceptive acts or practices, and acts or practices are unfair if they cause substantial injury to consumers that consumers cannot reasonably avoid themselves and that is not outweighed by countervailing benefits to consumers or competition. *Id.*

[23] *See, e.g.*, cases cited at footnote 16, *supra*.

[24] *See* Federal Trade Commission Roundtable, *Mobile Cramming, An FTC Roundtable* (May 8, 2013) (online at http://www.ftc.gov/news-events/events-calendar/2013/05/mobile-cramming-ftc-roundtable). The FCC also held a workshop on wireless cramming. *See* Federal Communications Commission Workshop, *Bill Shock and Cramming* (Apr. 17, 2013) (online at http://www.fcc.gov/events/workshop-bill-shock-and-cramming).

[25] 47 U.S.C. § 201(b).

[26] *See* Order, *FCC v. Assist 123, LLC*, at 3 (EB-TCD-12-00005541) (July 16, 2014) (online at http://www.fcc.gov/document/assist-123-pay-13m-resolve-wireless-cramming-investigation).

[27] 47 C.F.R. §§ 64.4200-64.2401.

[28] *Empowering Consumers to Prevent and Detect Billing for Unauthorized Charges ("Cramming"); Consumer Information and Disclosure; Truth-in-Billing and Billing Format*, 27 FCC Rcd 4436 (Apr. 27, 2012). The additional safeguards proposed regarding wireline cramming "require wireline carriers that currently offer blocking of third-party charges to clearly and conspicuously notify consumers of this option on their bills and websites, and at the point of sale; to place non-carrier third-party charges in a distinct bill section separate from all carrier charges; to provide subtotals in each section of the bill; and to display separate subtotals for carrier and non-carrier charges on the payment page of the bill." *Id.*

[29] *Consumer and Governmental Affairs Bureau Announces Comment Deadline for "Cramming" Further Notice of Proposed Rulemaking*, Public Notice, DA 12-833 (May 25, 2012).

[30] The comments were also joined by the National Consumer Law Center, Consumer Action, and the Center for Media Justice. *Comments of Center for Media Justice, Consumer Action, Consumer Federation of America, Consumers Union, National Consumer Law Center – On Behalf of its Low-Income Clients, and National Consumer League*, Federal Communications Commission, CG Docket No. 11-116, CG Docket No. 09-158, & CG Docket No. 98-170 (June 25, 2012).

[31] *Id.* at 18.

[32] *Id.* at 20.

[33] *Id.* at 20-21.

[34] *Id.* at 21-22.

[35] *Comments of CTIA-The Wireless Association*, Federal Communications Commission, CG Docket No. 11-116, CG Docket No. 09-158, & CG Docket No. 98-170 (June 25, 2012).

[36] *Consumer and Governmental Affairs Bureau Seeks to Refresh the Record Regarding "Cramming,"* CG Docket No. 11-116, CG Docket No 09-158, & CC Docket No. 98-170, Public Notice, DA 13-1807 (rel. Aug. 27, 2013). Issues on which FCC sought comment included the extent of cramming for consumers of wireline and wireless services, the need for an opt-in requirement and the mechanics of an opt-in process for wireline and wireless services, the details and efficacy of any other industry efforts to combat wireline and wireless cramming, whether different measures to combat cramming are appropriate for small and rural wireless carriers and other wireless carriers, and whether additional measures to combat wireline and wireless cramming are appropriate. *Id.* at 2-3.

[37] Senate Committee on Commerce, Science, and Transportation, *Rockefeller Asks Wireless Carriers for Information on Third-Party Charges* (June 12, 2012).

[38] Senate Committee on Commerce, Science, and Transportation, *Rockefeller Vows to Avert Wireless Cramming Scams on Consumers* (Mar. 1, 2013).

[39] Senate Committee on Commerce, Science, and Transportation, *Rockefeller Questions Billing Aggregators on Wireless Cramming* (Mar. 22, 2013).

[40] Senate Committee on Commerce, Science, and Transportation, *Senators Introduce Legislation to Stop Cramming on Telephone Bills* (June 12, 2013).

[41] Plaintiff's Original Petition, *Texas v. Mobile Messenger U.S. Inc., et al,* Tex. D. Ct., Travis County (Nov. 6, 2013).

[42] Letter from Chairman Rockefeller to Michael L. Iaccarino, Chief Executive Officer, Mobile Messenger (Nov. 26, 2013).

[43] Verizon Wireless, *Premium Messaging FAQs* (accessed July 27, 2014) (available at http://www.verizonwireless.com/support/faqs/Premium_TXT_and_MMS/faq_premium_txt_and_mms.html).

[44] *See, e.g.,* Letter from Chief Executive Officer, mBlox, to Senator John D. Rockefeller IV (Apr. 23, 2013); *VT. AG: 3 Firms End Extra Cellphone Bill Charges*, Associated Press (Nov. 21,

2013). *See also AT&T Mobility, Sprint and T-Mobile Will No Longer. . . . ,* Communications Daily (Nov. 25, 2013) (quoting Verizon General Counsel as saying that Verizon had "previously decided to exit the premium messaging business"). PSMS use continues for charitable giving and political contributions. Briefing by CTIA – The Wireless Association to Senate Commerce Committee Majority Staff (June 3, 2014); Briefing by Sprint Nextel to Senate Commerce Committee Majority Staff (July 16, 2014).

[45] *See, e.g.,* Master Services Agreement provided by Mobile Messenger to Senate Commerce Committee (AG-MM-COMM-001908-001911).

[46] Common shortcodes can also be used to allow consumers to make charitable donations and political contributions via text messaging. The Committee's inquiry focused on commercial shortcode charges.

[47] *See* Common Short Code Administration, *About Short Codes Frequently Asked Questions – CTIA Vetting* (online at http://www.usshortcodes.com/about-sms-short-codes/sms-marketingfaqs.php#.U8l0L6ggZss).

[48] *See, e.g.,* Letter from Executive Vice President, Federal Relations, AT&T, to Chairman John D. Rockefeller IV, at 5 (July 11, 2012); Letter from General Counsel, Verizon Wireless, to Chairman John D. Rockefeller IV, at 5 (July 11, 2012); Letter from Vice President, Federal Legislative Affairs, T-Mobile USA, to Chairman John D. Rockefeller IV, at 1 (July 11, 2012).

[49] Mobile Marketing Association, *Global Code of Conduct*; Mobile Marketing Association, *U.S. Consumer Best Practices for Messaging, Version 7.0* (Oct. 16, 2012) (online at: http://www.mmaglobal.com/files/bestpractices.pdf).

[50] *Id.*

[51] *Id.*

[52] Graphic was provided to the Committee by the company Boku.

[53] Mobile Marketing Association, *Global Code of Conduct* (July 15, 2008); Mobile Marketing Association, *U.S. Consumer Best Practices for Messaging, Version 7.0* (Oct. 16, 2012); and CTIA – The Wireless Association, *Mobile Commerce Compliance Handbook, Version 1.0* (June 4, 2012).

[54] Mobile Marketing Association, *Global Code of Conduct* at 1 (July 15, 2008); Mobile Marketing Association, *U.S. Consumer Best Practices for Messaging, Version 7.0* (Oct. 16, 2012). MMA defines "application provider" as an organization that offers network based software solutions. Mobile Marketing Association, *MMA Glossary – Application Provider* (2014) (online at http://mmaglobal.com/wiki/application-provider). "Publisher" is defined as a company that provides WAP sites [a website that is specifically designed and formatted for display on a mobile device] and/or facilitates the delivery of advertising via one or more WAP sites; also, as a publisher of mobile content, such as games and personalization products. Mobile Marketing Association, *MMA Glossary – Publisher* (2014) (online at http://mmaglobal.com/wiki/publisher).

[55] CTIA – The Wireless Association, *Consumer Bill of Rights* (July 1, 2013).

[56] CTIA has been screening all applicants for shortcodes by requiring basic identity and program information, such as the company name, corporate registration, and legal history. *See* Common Short Code Administration, *About Short Codes Frequently Asked Questions – CTIA Vetting* (online at http://www.usshortcodes.com/about-sms-short-codes/sms-marketingCTIA has worked with Aegis Mobile and WMC Global to conduct the vetting process. *See id.*; Aegis Mobile, *CTIA Vetting FAQ* (online at http://www.aegismobile.com/resources

[57] Briefing by CTIA – The Wireless Association to Senate Commerce Committee Majority Staff (June 3, 2014); Mobile Marketing Association, *U.S. Consumer Best Practices for Messaging, Version 7.0* at 23-24 (Oct. 16, 2012); CTIA – The Wireless Association, *Mobile Commerce Compliance Handbook, Version 1.0* at 3-4 (June 4, 2012). CTIA also included the same provisions in its updated Handbook. *See* CTIA – The Wireless Association, *Mobile Commerce Compliance Handbook, Version 1.2* at 5, 7 (Aug.1, 2013).

[58] *See CTIA – The Wireless Association Launches Common Short Codes Media Monitoring Process,* Business Wire (June 15, 2009) (online at http://www.businesswire.com/news/home/20090615005802/en/CTIA%E2%80%93The-Wireless- Association-Launches-Common-Short-Codes#.U8VyB6ggZss); WMC Global, *Frequently Asked Questions* (online at http://www.wmcglobal.com/faq.html); CTIA, *CTIA In-Market Monitoring Portal User Guide* (online at http://www.wmcglobal.com/assetsBriefing by CTIA – The Wireless Association to Senate Commerce Committee Majority Staff (June 3, 2014).

[59] Briefing by CTIA – The Wireless Association to Senate Commerce Committee Majority Staff (June 3, 2014); CTIA – The Wireless Association, *Mobile Commerce Compliance Handbook, Version 1.0,* at 6-7 (June 4, 2012).

[60] *Id.*

[61] Briefing by CTIA – The Wireless Association to Senate Commerce Committee Majority Staff (June 3, 2014).

[62] *See, e.g.,* WMC Global for CTIA – The Wireless Association, *In-Market Monitoring Update January 2011,* at 6 (Jan. 2011).

[63] Letter from Executive Vice President, Federal Relations, AT&T, to Chairman John D. Rockefeller IV, at 4 (July 11, 2012); Letter from General Counsel, Verizon Wireless, to Chairman John D. Rockefeller IV, at 6 (July 11, 2012); Letter from Vice President – Government Affairs, Sprint Nextel, to Chairman John D. Rockefeller IV, at 2 (July 11, 2012); Letter from Vice President, Federal Legislative Affairs, T- Mobile USA, to Chairman John D. Rockefeller IV, at 6 (July 11, 2012).

[64] *See, e.g.,* Sample Advanced Messaging Agreement for Marketing Messaging Hubs provided by mBlox to the Senate Commerce Committee (stating "At a minimum, programs shall be run in a manner that is congruous with the letter and spirit of the MMA Code of Conduct for Mobile Marketing") (000360).

[65] Letter from Assistant General Counsel, Verizon Wireless, to Senate Commerce Committee Majority Counsel, at 1-5 (July 12, 2013); Letter from Vice President – Government Affairs, Sprint Nextel, to Chairman John D. Rockefeller IV, at 3 (Mar. 22, 2013); Letter from Executive Vice President, Federal Relations, AT&T, to Chairman John D. Rockefeller IV, at 4-5 (July 11, 2012); Letter from Vice President, Federal Legislative Affairs, T-Mobile USA, to Chairman John D. Rockefeller IV, at 3 (July 11, 2012).

[66] Letter from Executive Vice President, Federal Relations, AT&T, to Chairman John D. Rockefeller IV, at 5 (July 11, 2012); Letter from General Counsel, Verizon Wireless, to Chairman John D. Rockefeller IV, Attachment A, at 6 (July 11, 2012); Letter from Vice President – Government Affairs, Sprint Nextel, to Chairman John D. Rockefeller IV, at 7 (July 11, 2012); Letter from Vice President, Federal Legislative Affairs, T-Mobile USA, to Chairman John D. Rockefeller IV, at 5 (July 11, 2012).

[67] One carrier stated that such audits are done "randomly" (Letter from Vice President – Government Affairs, Sprint Nextel, to Chairman John D. Rockefeller IV, at 8 (June 28, 2013)); while another stated they are done on at least a monthly basis (Letter from Assistant General Counsel, Verizon Wireless, to Senate Commerce Committee Majority Counsel, at 5-6 (July 12, 2013)).

[68] Letter from Vice President – Government Affairs, Sprint Nextel, to Chairman John D. Rockefeller IV, at 5-6 (June 28, 2013); Letter from Vice President, Federal Legislative Affairs, T-Mobile USA, to Chairman John D. Rockefeller IV, at 4-5 (June 28, 2013); Letter from Executive Vice President, Federal Relations, AT&T, to Chairman John D. Rockefeller IV, at 4-5 (July 2, 2013); Letter from Assistant General Counsel, Verizon Wireless, to Senate Commerce Committee Majority Counsel, at 8 (July 12, 2013).

[69] Letter from Vice President, Federal Legislative Affairs, T-Mobile USA, to Chairman John D. Rockefeller IV, at 6 (July 11, 2012); Letter from Vice President – Government Affairs, Sprint Nextel, to Chairman John D. Rockefeller IV, at 4 (June 28, 2013); Letter from Executive Vice President, Federal Relations, AT&T, to Chairman John D. Rockefeller IV, at 7 (July 2, 2013); Letter from Assistant General Counsel, Verizon Wireless, to Senate

Commerce Committee Majority Counsel, at 1 (July 12, 2013). Consumers are often made aware of these options at the time of purchase of a wireless plan, during customer service calls regarding the appearance of unauthorized charges on a bill, and on the carriers' websites.

[70] Letter from Vice President – Government Affairs, Sprint Nextel, to Chairman John D. Rockefeller IV, at 4 (June 28, 2013); Letter from Assistant General Counsel, Verizon Wireless, to Senate Commerce Committee Majority Counsel, at 6 (July 12, 2013); Update from AT&T, to Chairman John D. Rockefeller IV, at 2 (Mar. 11, 2013).

[71] Letter from Vice President, Federal Legislative Affairs, T-Mobile USA, to Chairman John D. Rockefeller IV, at 3 (July 11, 2012).

[72] Letter from Executive Vice President, Federal Relations, AT&T, to Chairman John D. Rockefeller IV, at 1 (July 11, 2012).

[73] Letter from Vice President – Government Affairs, Sprint Nextel, to Chairman John D. Rockefeller IV, at 2 (July, 11 2012).

[74] Letter from General Counsel, Verizon Wireless, to Chairman John D. Rockefeller IV, at 1 (July 11, 2012).

[75] Letter from Assistant General Counsel, Verizon Wireless, to Senate Commerce Committee Majority Counsel, at 10-12 (July 12, 2013); Letter from Vice President – Government Affairs, Sprint Nextel, to Chairman John D. Rockefeller IV, at 6-8 (June 28, 2013); Letter from Vice President, Federal Legislative Affairs, T-Mobile USA, to Chairman John D. Rockefeller IV, at 6-7 (June 28, 2013); and Letter from Executive Vice President, Federal Relations, AT&T, to Chairman John D. Rockefeller IV, at 7-11 (July 2, 2013).

[76] *See, e.g., Comments of CTIA – The Wireless Association*, Federal Communications Commission, CC Docket No. 98-170 (June 25, 2012); Commentary of Mike Altschul, General Counsel, CTIA – The Wireless Association, Federal Trade Commission, *Mobile Cramming, An FTC Roundtable* (May 8, 2013).

[77] *Comments of CTIA – The Wireless Association*, Federal Communications Commission, CC Docket No. 98-170, at 1-2 (June 25, 2012).

[78] Commentary of Cara Frey, General Counsel, Mobile Marketing Association, Federal Trade Commission, *Mobile Cramming, An FTC Roundtable* (May 8, 2013).

[79] Commentary of Jim Greenwell, Chief Executive Officer and President, BilltoMobile, Federal Trade Commission, *Mobile Cramming, An FTC Roundtable* (May 8, 2013) (estimating the volume of such billing to be between $2 to $3 billion).

[80] Letters from Carrier Representatives to Chairman John D. Rockefeller IV (July 2012).

[81] Four out of five aggregators provided revenues of content providers to the Committee. Letter from Head of Corporate Affairs and Communications, Ericsson Inc., to Chairman John D. Rockefeller IV (Apr. 19, 2013); Letter from Chief Administrative Officer and General Counsel, Motricity, Inc., to Chairman John D. Rockefeller IV (May 25, 2013); Letter from Attorney, Mobile Messenger, to Chairman John D. Rockefeller IV (Apr. 21, 2014); mBlox Response to Chairman John D. Rockefeller IV (Apr. 21, 2014).

[82] Response letters from carriers listed many aggregators. Letter from Executive Vice President, Federal Relations, AT&T, to Chairman John D. Rockefeller IV, at 3 (July 11, 2012); Letter from General Counsel, Verizon Wireless, to Chairman John D. Rockefeller IV, at Attachment A (July 11, 2012); Letter from Vice President – Government Affairs, Sprint Nextel, to Chairman John D. Rockefeller IV, at 4 (July 11, 2012); Letter from Vice President, Federal Legislative Affairs, T-Mobile USA, to Chairman John D. Rockefeller IV, at 1 (July 11, 2012).

[83] *See, e.g.,* Commentaries of Jim Greenwell, Chief Executive Officer and President, BilltoMobile, and Martine Niejadlik, Compliance Officer and Vice President of Customer Support, Boku, Federal Trade Commission, *Mobile Cramming, An FTC Roundtable* (May 8, 2013).

[84] Comments of California Public Utilities Commission, Federal Communications Commission, CC Docket No. 98-170, at 20 (Nov. 18, 2013).

[85] Committee staff reviewed a number of contracts between billing aggregators and wireless carriers that outlined the payment arrangements.

[86] Sample aggregator contract provided to the Senate Commerce Committee (000179).

[87] Sample aggregator contract provided to the Senate Commerce Committee (000066-000067).

[88] Between 2011 and 2013, carriers reported refunding $60,037,906 out of $495,134,687 in total wireless charges, including $25,095,834 in 2011, $23,250,885 in 2012, and $11,691,187 in 2013, with total billed, including $173,644,442 in 2011, $191,302,355 in 2012, and $130,187,888 in 2013. Comments of California Public Utilities Commission, Federal Communications Commission, CC Docket No. 98-170 (Nov. 18, 2013) and e-mail from CPUC Representatives to Senate Commerce Committee Majority Counsel (Apr. 23, 2014).

[89] *See Reply Comments of CTIA – The Wireless Association*, Federal Communications Commission, CC Docket No. 98-170, at 5 (Dec. 16, 2013) (arguing that "refund amounts and refund rates are flawed metrics for assessing instances of unauthorized charges on wireless bills. Carriers have consumer- friendly refund policies that cover a variety of situations and transactions – much more than just unauthorized third-party charges.").

[90] Commentary of Chris Witteman, Senior Staff Counsel, California Public Utilities Commission, Federal Trade Commission, *Mobile Cramming, An FTC Roundtable* (May 8, 2013).

[91] Letter from Consumer Affairs Branch, California Public Utilities Commission, to Senate Commerce Committee Majority Counsel (Jan. 31, 2013).

[92] Calculation of this percentage was based on the total number of wireless subscriber connections in California (34 million), and the United States (326.4 million) in 2012, as reported by CPUC and CTIA respectively. California Public Utilities Commission, *2012 Annual Report* (Feb. 1, 2013); CTIA – The Wireless Association, *Wireless Quick Facts* (last updated June 2014) (online at http://www.ctia.org/your- wireless-life/how-wireless-works/annual-wireless-industry-survey). Committee staff was unable to compare California and national wireless subscriber numbers for 2013, as CPUC's 2013 Annual Report did not include the number of wireless subscriber connections in California.

[93] Letter from T-Mobile USA, to Chairman John D. Rockefeller IV, at 5 (June 28, 2013).

[94] mBlox Response to Senate Commerce Committee (Mar. 25, 2014) (chart titled "US Cases (Jan-1-2012 – March-31-2013)").

[95] Federal Trade Commission, *Consumer Sentinel Network Data Book for January – December 2012*, at 84 (Feb. 2013) (showing 784 complaints for mobile unauthorized charges in 2010); Federal Trade Commission, *Consumer Sentinel Network Data Book for January – December 2013*, at 84 (Feb. 2014) (showing complaints for mobile unauthorized charges was 626 in 2011, 714 in 2012, and 363 in 2013).

[96] Center for Rural Studies at the University of Vermont, *Mobile Phone Third-Party Charge Authorization Study* (May 5, 2013). Following the release of survey results, CTIA engaged an expert to conduct an analysis of the Vermont Study. The analysis highlighted concerns with the methodology used for the study. The analysis was submitted to the Federal Trade Commission by CTIA on June 24, 2013.

[97] *See, e.g.,* Federal Communications Commission, CC Docket No. 98-170, Comments filed by AT&T, Inc. (July 20, 2012), (Dec. 16, 2013); T-Mobile USA Reply Comments (July 20, 2012); and Verizon Wireless (June 25, 2012).

[98] Opening Remarks by Commissioner Maureen Ohlhausen, Federal Trade Commission, *Mobile Cramming, An FTC Roundtable* (May 8, 2013).

[99] Sprint Nextel Response to Chairman John D. Rockefeller IV (Mar. 22, 2013).

[100] Complaint for Permanent Injunction and Other Equitable Relief, *Federal Trade Commission v. T- Mobile USA, Inc.*, W.D. Wa., at 4-9 (No. 2:14-cv-00967) (July 1, 2014).

[101] *Id.* at 5-6.

[102] *Id.* T-Mobile recently represented to Committee majority staff that the company has changed the way these charges are depicted in their wireless bills. Briefing by T-Mobile USA to Senate Commerce Majority Staff (July 17, 2014).

[103] *FTC v. T-Mobile, supra* note 100, at 6-9.

[104] Plaintiff's Original Verified Petition and Application for Ex Parte Temporary Restraining Order, Temporary Injunction, and Permanent Injunction, *Texas v. Eye Level Holdings, LLC, et al.*, Tex. D. Ct., Travis County (No. 1:11-cv-00178) (Mar. 11, 2011).

[105] Press Release, *Mobile Crammers Settle FTC Charges of Unauthorized Billing*, Federal Trade Commission (Nov. 21, 2013) (online at http://www.ftc.gov/news-events/press-releases/2013/11/mobile- crammers-settle-ftc-charges-unauthorized-billing).

[106] Press Release, *Operators of Massive Mobile Cramming Scheme Will Surrender More than $10 M in Assets in FTC Settlement*, Federal Trade Commission (June 13, 2014) (online at: http://www.ftc.gov/news-events/press-releases/2014/06/operators-massive-mobile-cramming-scheme- will-surrender-more-10m). The judgment was partially suspended based on defendants' inability to pay the full amount.

[107] Complaint for Permanent Injunction and Other Equitable Relief, *Federal Trade Commission v. T-Mobile USA, Inc.*, W.D. Wa. (No. 2:14-cv-00967) (July 1, 2014).

[108] Defendants named in state and federal enforcement action include: Jesta Digital, Bullroarer, Mobile Media Products, Bune, Wise Media, Tatto, Eye Level Holdings, Anacapa Media LLC, Tendenci Media, Bear Communications LLC, MDK Media, Mundo Media, SE Ventures, GMK Communications, MindKontrol Industries LLC, and Network One Commerce Inc. *Federal Trade Commission v. Jesta Digital, LLC, also d/b/a JAMSTER*, D.D.C. (No. 1:13-CV-01272) (Aug. 20, 2013); *Federal Trade Commission v. Tatto, et al.*, C.D. Cal (No. 2:13-cv13-8912-DSF-FFM) (Dec. 5, 2013); *Federal Trade Commission v. Wise Media, LLC, et al.*, N.D. Ga. (No. 1:13cv1234) (Apr. 16, 2013); *Texas v. Eye Level Holdings, LLC, et al.*, Tex. D. Ct., Travis County (No. 1:11-cv-00178) (Mar. 11, 2011); *Texas v. Mobile Messenger U.S. Inc., et al*, Tex. D. Ct., Travis County (Nov. 6, 2013); *Federal Trade Commission v. MDK Media, Inc., et al.*, C.D. Cal (No. 2:14-cv-05099-JFW-SH) (July 3, 2014).

[109] *AT&T Settles with Florida AG Over Mobile Content Ads*, Mobile Marketer (Mar. 3, 2008); *T-Mobile $600k Settlement with Florida AG Affects All Mobile Content Marketing,* Mobile Marketer (July 22, 2010); *Sprint Settles Cell Phone Cramming Charges in Florida*, Consumer Affairs (Oct. 8, 2010).

[110] *See, e.g., Verizon Settlement with Florida AG Affects All Marketing of Mobile Content*, Mobile Marketer (June 29, 2009); *In the Matter of Verizon Wireless LLC and Alltel Communications, LLC*, Assurance of Voluntary Compliance at 5-13 (June 16, 2009). Specifically, the settlements resulting from the Florida actions require that certain provisions be included in the contracts between the carriers and any companies that "advertise, aggregate billing for, offer and/or sell mobile content." These include: (1) a prohibition on using words like "free," "complimentary," "without charge" or other similar terms without clear and conspicuous disclosure that the consumer will have to pay for a subscription in order to receive the content; (2) specifications for font size and color on all consumer disclosures for web-based advertising for mobile content; and (3) certain price and billing disclosures must be made "above the fold" on the mobile "submit" and "PIN submit" pages. In addition to these best practices, the settlements also required the carriers to establish monetary compensation programs for consumers who had experienced unauthorized third-party billing charges. *Id.* at 4-10, 13-14.

[111] Briefing by CTIA – The Wireless Association to Senate Commerce Committee Majority Staff (June 3, 2014).

[112] *See* Part III.B.1, *supra*.

[113] WMC Global for CTIA – The Wireless Association, *In-Market Monitoring Update January 2011*, at 2 (Jan. 2011). According to the report, 18,304 offers were tested. *Id.* at 3. In May 2011, the monthly in-market auditing reports began including specific breakdowns of premium messaging testing results and standard rate testing results, but the January 2011 report does not provide that breakdown. CTIA has represented to the Committee that the PSMS testing in January 2011 had a failure rate of 97%.

[114] The January 2011 In-Market Monitoring Update showed that 99.57% of interceptions failed at severity level 1. *Id.* at 3.

[115] CTIA – The Wireless Association, *Mobile Commerce Compliance Handbook, Version 1.0*, at 6 (June 4, 2012). For example, the most common of the January 2011 violations concerned the "no account holder authorization disclosure" requirement concerning how to disclose that the account holder must authorize purchases. *In-Market Monitoring Update January 2011,* at 2 (Jan. 2011); CTIA – The Wireless Association, *Mobile Commerce Compliance Handbook, Version 1.0* at 8 (June 4, 2012) (describing standards).

[116] According to the September 2011 report, 49% of the interceptions failed while 51% passed. WMC Global for CTIA – The Wireless Association, *In-Market Monitoring Update September 2011,* at 3 (Sept. 2011).

[117] WMC Global for CTIA – The Wireless Association, *In-Market Monitoring Update January 2012*, at 3 (Jan. 2012).

[118] *See, e.g.*, WMC Global for CTIA – The Wireless Association, *In-Market Monitoring Update September 2012,* at 3 (Sept. 2012) (showing 96% of interceptions passed); WMC Global for CTIA– The Wireless Association, *In-Market Monitoring Update October 2012,* at 3 (Oct. 2012) (showing 97% of interceptions passed), WMC Global for CTIA– The Wireless Association, *In-Market Monitoring Update December 2012,* at 3 (Dec. 2012) (showing 98% of interceptions passed).

[119] For example, similar to the policies described above, for third-party billing on landline phones, phone companies instituted policies providing for screening of vendors, the option for consumers to block third- party billing, and customer complaint thresholds that trigger corrective action. For a detailed discussion of industry self-regulation initiatives to address cramming on wireline phone bills *see* Senate Committee on Commerce, Science, and Transportation, *Staff Report on Unauthorized Third-Party Charges on Telephone Bills*, at 30-33 (July 12, 2011).

[120] *Id.* at 4-5.

[121] *Comments of CTIA – The Wireless Association*, Federal Communications Commission, CC Docket No. 98-170, at 2 (June 25, 2012); Letter from General Counsel, Verizon Wireless, to Chairman John D. Rockefeller IV, at 2 (July 11, 2012) (noting there is no analogue to the double opt-in in the wireline billing context).

[122] Commentary of Mike Altschul, General Counsel, CTIA – The Wireless Association, Federal Trade Commission, *Mobile Cramming, An FTC Roundtable* (May 8, 2013).

[123] Complaint for Permanent Injunction and Other Equitable Relief, *Federal Trade Commission v. Tatto, et al.*, C.D. Cal (No. 2:13-cv13-8912-DSF-FFM) (Dec. 5, 2013).

[124] Complaint for Permanent Injunction and Other Equitable Relief and Exhibits, *Federal Trade Commission v. Wise Media, LLC, et al.*, N.D. Ga. (No. 1:13cv1234) (Apr. 16, 2013).

[125] Plaintiff's Original Verified Petition and Application for Ex Parte Temporary Restraining Order, Temporary Injunction, and Permanent Injunction, *Texas v. Eye Level Holdings, LLC, et al.*, Tex. D. Ct., Travis County (No. 1:11-cv-00178) (Mar. 11, 2011). The case settled in 2012.

[126] Plaintiff's Original Petition, *Texas v. Mobile Messenger U.S. Inc., et al.,* Tex. D. Ct., Travis County, at 12 (Nov. 6, 2013).

[127] *Id.*

[128] *Id.* at 14-15.

[129] *Id.* at 15.

[130] *Id.* at 18-22.

[131] *See, e.g.,* Commentary of Mike Altschul, General Counsel, CTIA – The Wireless Association, Federal Trade Commission, *Mobile Cramming, An FTC Roundtable* (May 8, 2013) (stating that the negative option, where companies would instruct the consumer to reply "stop" or be charged, in the double opt-in process, which was utilized by many of the subjects in law enforcement proceedings, was "not compliant with the industry best practices" and use of this negative option was "not playing by the rules"). *See also* Commentary of John Bruner,

Chief Operating Officer, Aegis Mobile, Federal Trade Commission, *Mobile Cramming, An FTC Roundtable* (May 8, 2013) (stating, "[W]hat we see in the market is not a violation of the double opt-in where it's being skipped necessarily. What we usually see is that consumers are either, through stacked marketing or deceptive advertising, double opting in and not realizing that they had purchased something. And, so, you know, the process, the physical process itself seems to be a very sound process for purchase. It's more the method leading up to getting a consumer to perform that function.").

[132] *See* Section IV.B.1 for discussion of industry and consumer advocate views on the significance of refund rates.

[133] Mobile Messenger Subpoena Response to Chairman John D. Rockefeller IV (Mar. 31, 2013), (Apr. 21, 2014), (Apr. 22, 2014). Because documents produced to the Committee concerned a small number of vendors, findings of this review provide a sample rather than a comprehensive review of carrier practices, and a review of communications relating to other vendors would be necessary to draw broad conclusions about an individual carrier's practices generally.

[134] Letter from Assistant General Counsel of Verizon Wireless, to Senate Commerce Committee Majority Counsel, at 8 (July 12, 2013).

[135] E-mail from Mobile Messenger Sales Employee to Mobile Messenger Account Manager (May 30, 2013) (with subject line: "05/29/2013 Refund Report for AT&T/Sprint/T-Mobile/VZW (Anacapa)") (AG-MM-COMM-043461-043464).

[136] *Id.*

[137] Letter from Executive Vice President, AT&T, to Chairman John D. Rockefeller IV, at 4 (July 2, 2013).

[138] *Id.*

[139] E-mail from Mobile Messenger Sales Employee to Mobile Messenger Account Manager (May 30, 2013) (with subject line: "05/29/2013 Refund Report for AT&T/Sprint/T-Mobile/VZW (Anacapa)") (AG-MM-COMM-043461-043464).

[140] Letter from Vice President, Government Affairs, Sprint Nextel, to Chairman John D. Rockefeller IV, at 5 (June 28, 2013).

[141] E-mail from Mobile Messenger Sales Employee to Mobile Messenger Account Manager (May 30, 2013) (with subject line: "05/29/2013 Refund Report for AT&T/Sprint/T-Mobile/VZW (Anacapa)") (AG-MM-COMM-043461-043464).

[142] Letter from Vice President, Federal Legislative Affairs, T-Mobile USA, to Chairman John D. Rockefeller IV, at 5 (June 28, 2013).

[143] E-mail from Mobile Messenger Sales Employee to Mobile Messenger Account Manager (May 30, 2013) (with subject line: "05/29/2013 Refund Report for AT&T/Sprint/T-Mobile/VZW (Anacapa)") (AG-MM-COMM-043461-043464).

[144] E-mail from Mobile Messenger Account Manager to Vendor Employee (Aug. 4, 2011) (with subject line: "FW: Client Alert – Carrier Alert: T-Mobile – Modifications to Their Refund Performance Improvement Plan") (AG-MM-COMM-016777-016779).

[145] *Id.*

[146] Briefing by VISA Representatives to Senate Commerce Committee Majority Staff (July 10, 2014).

[147] *See, e.g.,* Letter from Verizon Wireless Director to Mobile Messenger Compliance & Consumer Protection Employee (Oct. 20, 2011) (AG-MM-COMM-030220-22) (describing that Ontario Corp. had "repeatedly violated the requirements applicable to premium messaging campaigns" and "given the repeated and serious nature of the violations, Verizon Wireless ha[d] decided that all premium messaging campaigns managed by the content provider must be terminated"); Spreadsheet created by Mobile Messenger listing status on several shortcodes of Sprint (last saved Aug. 10, 2011) (AG-MM-COMM-021051) (noting 5 terminated codes and several codes that were temporarily suspended); E-mail from AT&T Senior Account Manager to Mobile Messenger Employees (Jan. 14, 2013) (AG-MM-COMM- 125203-4) (noting AT&T termination of all short codes associated with AVL

marketing); E-mail from T- Mobile Compliance to Mobile Messenger Employees (Oct. 4, 2013) (AG-MM-COMM-143991) (noting 3 shortcodes that received 3 strikes under T-Mobile's PIP policy and directing Mobile Messenger employees to "immediately terminate all billing and related services currently operating" and the short codes).

[148] E-mail from WMC Global AT&T Account Manager to Mobile Messenger Employees (Oct. 15, 2013) (AG-MM-COMM-057077-057080).

[149] E-mail from AT&T Mobility Marketing Manager to Mobile Messenger Employees (Oct. 16, 2013) (AG-MM-COMM-056884-056885).

[150] *Id.*

[151] *Id.*

[152] E-mail from Mobile Messenger Compliance & Analytics to NeoImage Employees (July 5, 2011) (AG- MM-COMM-112226-112227) (the 11 shortcode refunds totaled $457,252.29). As indicated by this e-mail and other Mobile Messenger documents, NeoImage personnel had e-mail addresses at "mundomedia.com." *See also* Mobile Messenger Spreadsheet Response to Subpoena Item 2a and 2c (AG-MM-COMM-001128) (Apr. 21, 2014) (listing company directors with email addresses, including NeoImage personnel with mundomedia.com addresses).

[153] E-mail from Mobile Messenger Account Management Employee to NeoImage Employees (Oct. 9, 2012) (AG-MM-COMM-585462-585463) (the 11 shortcode refunds totaled $594,479).

[154] *See, e.g.,* E-mail notification from Mobile Messenger Compliance & Analytics to NeoImage Employees (Oct. 4, 2011) (AG-MM-COMM-024918-024919) (noting that 8 of their shortcodes violated Sprint's threshold, with the highest rate at 28.79% and the refunds for all 8 totaling over $600,000 for the three-month period); E-mail from Mobile Messenger Compliance & Analytics to NeoImage Employees (June 7, 2011) (AG-MM-COMM-099369) (indicating that 6 shortcodes had refund rates exceeding Verizon's thresholds in May, with the highest rate reported as 22.23% and refunds for all 6 totaling over $340,000).

[155] E-mail from Mobile Messenger Compliance Analytics to NeoImage Employees (Mar. 9, 2012) (AG- MM-COMM-590211-590212).

[156] E-mail from Mobile Messenger Compliance Analytics to NeoImage Employees (Apr. 2, 2012) (AG- MM-COMM-584986-584987).

[157] E-mail from Mobile Messenger Compliance Analytics to NeoImage Employees (June 5, 2012) (AG- MM-COMM-568009-568010). The Mobile Messenger document production did not appear to include an AT&T excess refund rate notification for the month of April 2012.

[158] E-mail from Mobile Messenger Compliance Analytics to NeoImage Employees (Jan. 11, 2011) (AG- MM-COMM-060591).

[159] E-mail from Mobile Messenger Compliance Analytics to NeoImage Employees (Mar. 3, 2011) (AG- MM-COMM-146312-146313).

[160] E-mail from Mobile Messenger Compliance Analytics to NeoImage Employees (Apr. 4, 2011) (AG- MM-COMM-220607-220608).

[161] Letter from Verizon Wireless Director to Mobile Messenger Compliance & Consumer Protection Employee (Oct. 20, 2011) (AG-MM-COMM-032198-032199) (listing shortcodes 91097, 33999, 72449, 40684, 25692, 89147, 88922, 21500, 86358, 56255, 53405, and 62131).

[162] E-mail from Mobile Messenger Compliance & Analytics to NeoImage Employees, Subject: Notice: AT&T Refund Ratio Exceeded (Mar. 2, 2011) (AG-MM-COMM-145222-145223) (showing shortcode 89147 February refund rate was 45.52%); E-mail from Mobile Messenger Compliance & Analytics to NeoImage Employees, Subject: Notice: AT&T Refund Ratio Exceeded (Apr. 4, 2011) (AG-MM-COMM- 220637-220638) (showing shortcode 91097 March refund rate was 50.65%); E-mail from Mobile Messenger Compliance & Analytics to NeoImage Employees, Subject: Notice: Sprint Refund Ratio Exceeded (Aug. 1, 2011) (AG-MM-COMM-012239) (noting shortcode 53405 April 2011-June 2011 refund ratio was 31.67% and 56255 was 15.95%); E-mail from Mobile

Messenger Compliance & Analytics to NeoImage Employees, Subject: Notice: Verizon Refund Ratio Exceeded – June 2011 (July 5, 2011) (AG-MM-COMM-112223-112224) (noting June refund rate for 56255 was 10.74%, 33999 was 15.52%, 86358 was 11.8%, and 88922 was 11.15%); E-mail from Mobile Messenger Compliance & Analytics to NeoImage Employees, Subject: Notice: T-Mobile Refund Ratio Exceeded (Sept. 2, 2011) (AG-MM-COMM-290976-290977) (noting shortcode 33999 August refund rate was 19.05%).

[163] Letter from Verizon Wireless Director to Mobile Messenger Compliance & Consumer Protection Employee (Oct. 20, 2011) (AG-MM-COMM-032198-032199).

[164] *Id.*

[165] *See, e.g.*, Mobile Messenger, *Settlement Statement for NeoImage December 1, 2011-December 31, 2011*, at 10 (Jan. 2012) (showing a $11.25 balance due for Verizon Wireless); Mobile Messenger, *Settlement Statement for NeoImage January 1, 2012-January 31, 2012*, at 10 (Feb. 2012) (showing a $10.75 due for Verizon Wireless); Mobile Messenger, *Settlement Statement for NeoImage March 1, 2012- March 31, 2012*, at 7 (Apr. 2012) (Verizon does not appear on the statement).

[166] Mobile Messenger, *Settlement Statement for NeoImage Jan. 1, 2012-Jan. 31, 2012* (Feb. 2012) (AG- MM-042772-042781); Mobile Messenger, *Settlement Statement for NeoImage Mar. 1, 2012-Mar. 31, 2012* (Apr. 2012) (AG-MM-585109-585115); Mobile Messenger, *Settlement Statement for NeoImage Apr. 1, 2012-Apr. 31, 2012* (May. 2012) (AG-MM-562235-562241); Mobile Messenger, *Settlement Statement for NeoImage May 1, 2012-May 31, 2012* (June. 2012) (AG-MM-568311-568318); Mobile Messenger, *Settlement Statement for NeoImage June 1, 2012-June 31, 2012* (July. 2012) (AG-MM- 563935-563941); Mobile Messenger, *Settlement Statement for NeoImage July 1, 2012-Aug. 31, 2012* (Sept. 2012) (AG-MM-588084-588098); Mobile Messenger, *Settlement Statement for NeoImage Sept. 1, 2012-Sept. 31, 2012* (Oct. 2012) (AG-MM-192132-102137); Mobile Messenger, *Settlement Statement for NeoImage Dec. 1, 2012-Dec. 31, 2012* (Jan. 2013) (AG-MM-591730-591735) (Mobile Messenger's production to Committee staff was missing settlement statements for several months for 2012 so NeoImage's total charge of $10 million is a conservative total).

[167] Mobile Messenger, *Settlement Statement for NeoImage Jan. 1, 2012-Jan. 31, 2012*, at 3, 5, 7 (Feb. 2012).

[168] Mobile Messenger, *Settlement Statement for NeoImage Mar. 1, 2012-Mar. 31, 2012*, at 2, 3, 4 (Apr. 2012).

[169] E-mail from Mobile Messenger Compliance Analytics to NeoImage Employees (Feb. 2, 2012) (with subject line: Notice: T-Mobile Refund Ratio Exceeded) (AG-MM-COMM-040764-040765).

[170] E-mail from Mobile Messenger Compliance Analytics to NeoImage Employees (Feb. 2, 2012) (with subject line: Notice: AT&T Refund Ratio Exceeded) (AG-MM-COMM-040831-040832).

[171] Letter from Vice President – Government Affairs, Sprint Nextel, to Chairman John D. Rockefeller IV, at 7 (July 11, 2012).

[172] Letter from Executive Vice President, Federal Relations, AT&T, to Chairman John D. Rockefeller IV, at 7 (July 2, 2013).

[173] Letter from Vice President, Federal Legislative Affairs, T-Mobile USA, to Chairman John D. Rockefeller IV, at 3 (June 28, 2013).

[174] Mobile Messenger Narrative Response to Chairman John D. Rockefeller IV (May 24, 2013) (MM Confidential 000004, 000050).

[175] Mobile Messenger Narrative Response to Chairman John D. Rockefeller IV (May 24, 2013) (MM Confidential 000050).

[176] Plaintiff's Original Petition, *Texas v. Mobile Messenger U.S. Inc., et al.*, Tex. D. Ct., Travis County (Nov. 6, 2013). This case remains open at the time of this report.

[177] *Id.*

[178] *See, e.g.*, E-mail from Mobile Messenger Account Manager to NeoImages Employees (Aug. 24, 2011) (AG-MM-COMM-022790-022795) and E-mail from Mobile Messenger Sales

Employee to Mobile Messenger Account Manager (May 30, 2013) (AG-MM-COMM-043461-043464) (cataloguing refund rates across major carriers for different vendors); Letter from Verizon Wireless Director to Mobile Messenger Compliance & Consumer Protection Employee, Re: Urgent Resolution of Violations (Oct. 20, 2011).

[179] See discussion *supra* at Section IV.D.2.c; *see also* Assignment of Rights and Amendment Among Neo Images, Inc., and Subscriber Management Services, LLC and Mobile Messenger US, Inc. (signed March 19, 2012) (AG-MM-COMM 001964-2031).

[180] Some carriers still support PSMS billing for charitable donations and political contributions. Briefing by Verizon Wireless to Senate Commerce Committee Majority Staff (July 16, 2014); Briefing by Sprint Nextel to Senate Commerce Committee Majority Staff (July 16, 2014); Briefing by T-Mobile USA to Senate Commerce Majority Staff (July 17, 2014).

[181] *Comments of CTIA – The Wireless Association*, Federal Communications Commission, CC Docket No. 98-170, at 4 (Nov. 18, 2013).

[182] *Id.* at 3 (citing *Study: Popularity of Direct Carrier Billing on the Rise*, Mobile Payments Today (Sept. 4, 2012).

[183] *Id.* at 4-5.

[184] Briefing by Verizon Wireless to Senate Commerce Committee Majority Staff (July 16, 2014).

[185] Briefing by Google to Senate Commerce Committee Majority Staff (July 11, 2014); Briefing by Verizon Wireless to Senate Commerce Committee Majority Staff (July 16, 2014); Briefing by Sprint Nextel to Senate Commerce Committee Majority Staff (July 16, 2014); Briefing by T-Mobile USA to Senate Commerce Committee Majority Staff (July 17, 2014).

[186] Briefing by Google to Senate Commerce Committee Majority Staff (July 11, 2014).

[187] Briefing by Boku to Senate Commerce Committee Majority Staff (June 23, 2014); Briefing by BilltoMobile to Senate Commerce Committee Majority Staff (Feb. 24, 2014); Briefing by Verizon Wireless to Senate Commerce Committee Majority Staff (July 16, 2014); Briefing by Sprint Nextel to Senate Commerce Committee Majority Staff (July 16, 2014); Briefing by T-Mobile USA to Senate Commerce Committee Majority Staff (July 17, 2014).

[188] Examples include, among others, Boku, a San Francisco-based company, and BilltoMobile, a San Jose-based company, both of which contract with major U.S. wireless carriers. *See* BilltoMobile, Home Page (online at http://www.billtomobile.com/); Briefing by BilltoMobile to Senate Commerce Committee Majority Staff (Feb. 24, 2014); Boku, Home Page (online at http://www.boku.com/); Briefing by Boku to Senate Commerce Committee Majority Staff (June 23, 2014).

[189] Briefing by CTIA – The Wireless Association to Senate Commerce Committee Majority Staff (June 3, 2014); Briefing by Sprint Nextel to Senate Commerce Committee Majority Staff (July 16, 2014); Briefing by Verizon Wireless to Senate Commerce Committee Majority Staff (July 16, 2014); Briefing by T- Mobile USA to Senate Commerce Committee Majority Staff (July 17, 2014).

[190] Briefing by Sprint Nextel to Senate Commerce Committee Majority Staff (July 16, 2014); Briefing by Verizon Wireless to Senate Commerce Committee Majority Staff (July 16, 2014).

[191] Briefing by Boku to Senate Commerce Committee Majority Staff (June 23, 2014); Briefing by Federal Reserve Bank to Senate Commerce Committee Majority Staff (July 18, 2014).

[192] *See* n. 181, *supra*.

[193] Briefing by Boku to Senate Commerce Committee Majority Staff (June 23, 2014); Briefing by Federal Reserve Bank to Senate Commerce Committee Majority Staff (July 18, 2014).

[194] Briefing by Sprint Nextel to Senate Commerce Committee Majority Staff (July 16, 2014); Briefing by Verizon Wireless to Senate Commerce Committee Majority Staff (July 16, 2014); Briefing by T-Mobile USA to Senate Commerce Committee Majority Staff (July 17, 2014).

In: Cramming and Third Party Billing … ISBN: 978-1-63463-121-1
Editor: Christopher Rye

Chapter 2

MOBILE CRAMMING: AN FTC STAFF REPORT*

Federal Trade Commission

EXECUTIVE SUMMARY

The widespread adoption of mobile devices has provided many important benefits to consumers, including the convenience of paying for goods and services using a phone. One mobile payment option is known as "carrier billing" – the ability to charge a good or service directly to a mobile phone account. Industry observers have noted the potential usefulness of carrier billing to consumers, and the industry continues to devote resources to promoting carrier billing as a payment option. As stakeholders have noted, carrier billing of third-party charges may be particularly beneficial for unbanked and underbanked consumers. Additionally, consumers have used text messages to donate funds to a charitable organization, with the charge placed on their mobile phone account.

As carrier billing has developed, however, fraud has become a significant problem for consumers. In particular, mobile cramming – the unlawful practice of placing unauthorized third-party charges on mobile phone accounts – is a significant concern. Mobile cramming occurs when consumers are signed up and billed for third-party services, such as ringtones and recurring

* This is an edited, reformatted and augmented version of a report issued July 2014.

text messages containing trivia or horoscopes, without their knowledge or consent. In six recent enforcement actions, the Commission has alleged that such practices have cost consumers many millions of dollars, and in just three of these actions, defendants have agreed to orders imposing judgments totaling more than $160 million.

As part of its consumer protection mission, the Federal Trade Commission ("FTC" or "Commission") works to ensure that consumer protections keep pace with developing technologies and payment mechanisms. In late 2013, following a Commission roundtable on mobile cramming and a number of FTC and state enforcement actions highlighting the prevalence of mobile cramming, the four largest mobile carriers stated that they would discontinue one form of carrier billing for commercial transactions – "Premium SMS" billing. Market participants continue to implement other kinds of carrier billing arrangements, however. In doing so, it is imperative that they keep in mind the same fundamental consumer protection principles that apply regardless of the type of technology used for billing. All stakeholders have an interest in combating cramming on the carrier billing platform, as promoting consumer trust is integral to promoting more widespread adoption of mobile payment systems and further innovation.

Based on evidence reviewed to date, FTC staff recommends certain best practices for industry participants to protect consumers against unwanted charges while enabling innovation and consumer access to another payment mechanism.

First, mobile carriers should give consumers the option to block all third-party charges on their phone accounts. At activation, carriers should inform consumers that third-party charges may be placed on their mobile accounts and carriers should give consumers the opportunity to block all charges at that time. Carriers should also clearly and prominently inform consumers of options to block charges from third parties while accounts are active, including on the carriers' websites. Additionally, carriers should consider offering consumers the ability to block or allow only specific providers, or to block commercial providers only.

Second, advertisements for products or services charged to a mobile bill must not be deceptive. Merchants are in the first instance responsible for ensuring that their practices – including any advertising, marketing, and opt-in processes – are not deceptive. Information about price should be disclosed clearly and conspicuously before charging a consumer's mobile account. Further, carriers should implement reasonable procedures to scrutinize risky or suspicious merchants and terminate or take other appropriate steps against companies engaging in unlawful practices. Such scrutiny should also be applied if a carrier becomes aware that a merchant has run an earlier campaign containing deceptive advertising or engaged in unauthorized billing on landline phones.

Third, it is critical that consumers provide their express, informed consent to charges before they are billed to a mobile account, and that reliable records of such authorizations are maintained. The unreliability of many merchants' claims that they have obtained consumer consent suggests that more centralized control by carriers and intermediaries of the consumer opt-in process and authorization records is needed. Mobile carriers should implement policies, or strengthen existing polices, to investigate and take appropriate action when consumer refund requests, complaints, or other factors indicate that a merchant may be cramming charges without consumers' consent.

Fourth, all charges for third-party services should be clearly and conspicuously disclosed to consumers in a non-deceptive manner. In particular, on a phone bill, the name of the service and any associated bill heading should relate to the product offered and not suggest an affiliation with the carrier's service. Carriers should consider ways to make third-party charges more conspicuous, such as by providing separate subtotals for carrier and third-party charges wherever total charges are disclosed. Carriers also should consider whether consumers who auto-pay their bills would benefit from receiving a separate notification of third-party charges. Further, consumers with prepaid phone plans who do not typically receive phone bills should receive such a notification from the carrier.

Fifth, carriers should implement an effective dispute resolution process. Such a process should be clear and consistent and enable consumers to dispute suspicious charges on their mobile accounts and obtain refunds for unauthorized charges. Consumers have reported difficulties and inconsistent experiences with carriers' dispute resolution policies for third-party charges. Further, given the extensive evidence that consumers are often unaware of third-party charges on their phone bills, carriers should grant consumer refund requests for recurring unauthorized charges that the carrier concludes were crammed, including refunds for the same recurring charge in previous months to the extent it is practicable to identify those prior charges. When a carrier terminates a third party's billing activities due to unauthorized charges, the carrier should notify consumers who incurred charges from the third party to allow them to request a refund.

I. INTRODUCTION

In recent years, the use of mobile devices has grown so rapidly that mobile devices now outnumber people in the United States, and nearly as many people have a mobile phone as have a bank account.[1] Given these facts, it is no surprise that companies are increasingly offering consumers ways to charge payments for third-party goods and services to their mobile phone accounts – a payment method often referred to as "mobile carrier billing" or "carrier billing." Indeed, carrier billing has been estimated to be the most popular mobile payment system in use in the world today.[2] This payment method offers many potential benefits for consumers who want to use their mobile phones to pay for goods and services and who prefer that the charges be placed on their phone bill rather than on a credit or debit card, for example. In fact, carrier billing may be especially beneficial for unbanked and underbanked consumers.

Carrier billing, however, can also raise concerns that the practice will be misused to charge consumers for items without their consent. This concern has been borne out in consumer complaints and other evidence indicating that in many cases consumers have not authorized charges that have been billed to their phone accounts. Unauthorized third-party charges on mobile phone accounts – a practice known as "mobile cramming" – occurs when consumers are enrolled and billed for third-party services, such as ringtones and recurring text messages containing trivia or horoscopes, without the consumers' knowledge or consent. Third parties either obtain consumers' phone numbers and falsely claim that the consumers have signed up for services and authorized charges, or use deceptive means to obtain consumers' mobile phone numbers – such as by offering free prizes – and then begin charging consumers' phone accounts for recurring third-party charges for purported services unrelated to the offer. These charges often are difficult to locate in phone bills, and many consumers do not notice them or do not understand that such charges are associated with an item unrelated to their phone service. Further, many consumers do not even receive the supposed services for which they are being charged. Mobile cramming thus harms consumers and also undermines the use of carrier billing as a legitimate payment option. As carrier billing is an emerging payment method being actively promoted by carriers, it is in the interest of all stakeholders to take proactive steps now to ensure its reliability and foster consumer trust to enable the payment method to reach its full potential.

For the past two decades, one of the top priorities at the FTC has been ensuring that consumer protections keep pace with emerging technologies, including mobile technologies. Among other things, the FTC has brought enforcement actions to combat mobile cramming and provide restitution to injured consumers,[3] recommended the adoption of certain baseline consumer protections,[4] and encouraged public dialogue among industry stakeholders to identify potential solutions. To advance that dialogue, on May 8, 2013, the FTC convened a roundtable of industry participants, consumer advocates, regulators, and other interested parties to discuss and gather information about potential approaches to combat mobile cramming (the "roundtable").[5]

This FTC staff report addresses mobile carrier billing and summarizes information from the roundtable and public comments submitted in connection with the roundtable; data obtained from public sources; and evidence from the FTC's enforcement actions as well as multiple enforcement actions brought by states. It also recommends certain best practices for industry participants to protect consumers against mobile cramming. Specifically, Part II of this report describes mobile third-party billing, including the types of entities involved in the industry. Part III describes mobile cramming and its prevalence. Part IV describes current strategies used to combat mobile cramming and FTC staff's views on whether additional action is needed.

II. Mobile Third-Party Billing

This section describes the potential uses of carrier billing, the types of carrier billing, and the different market participants involved.

A. Uses of Mobile Carrier Billing

As discussed in the FTC's March 2013 staff report on mobile payments and at the roundtable, major phone carriers permit consumers to charge payments for third-party goods and services directly to their mobile phone accounts, which is known generally as "mobile carrier billing" or just "carrier billing,"[6] as an alternative to paying for an item with a credit or debit card, for example.

Roundtable participants and other industry observers have noted the potential usefulness of mobile carrier billing to consumers. Carrier billing of third-party charges may be useful for consumers who do not have credit cards, or do not want to use them, especially for small transactions. In this way, carrier billing may be beneficial for unbanked and underbanked consumers.[7] One roundtable participant also noted that this payment channel may be an appealing way of transacting for members of the millennial generation, who have grown up in a mobile-centric world and may often use a mobile phone as a primary way to make electronic payments.[8]

In the commercial context, consumers have been able to use carrier billing to pay for such items as text message-based subscription services and digital goods such as virtual currency in a mobile game app. In the charitable context, consumers have used text messages to donate funds to a charitable organization, such as in the aftermath of a natural disaster, with the charge placed on their mobile phone bill.[9] Additionally, consumers have made political contributions by sending text messages and having the charges placed on their mobile phone bills.[10]

B. Types of Carrier Billing Arrangements

Historically, third-party charges have often involved a text-messaging component, whereby a consumer purportedly authorizes charges by texting a particular five or six-digit number known as a "short code." Typically, in authorized transactions, the consumer initiates a transaction by sending a text message to a short code or entering his or her phone number on a website. The consumer then receives a text message with additional information about the good or service offered, and the consumer responds to confirm the charge, such as by sending a confirmatory response text. This type of mobile carrier billing is called "Premium SMS" billing, as text messages are also known as "SMS" ("short message service") messages.

Since the adoption of smartphones with advanced mobile web browsing capabilities and the more widespread use of mobile apps, there has been an increasing use of other forms of carrier billing arrangements, known as "direct carrier billing" (or "DCB") arrangements. In DCB arrangements, a consumer does not necessarily need to send or receive a text message to initiate or complete a transaction that is billed to a mobile account. Instead, a consumer can initiate a transaction on a mobile website or within a mobile app, and the merchant can have the charge placed on the consumer's mobile account through back-end arrangements that involve the mobile carriers.

In the past few years, commercial Premium SMS billing has generated billions of dollars of revenue, and one industry participant estimated that third-party charges on mobile bills constituted a $2 to $3 billion dollar annual market.[11] Several roundtable panelists, however, suggested that the Premium SMS market had been declining in size over the last few years, and that one key reason for this is the emergence of mobile apps, which allow consumers to obtain content (such as ringtones) that they may have previously purchased using Premium SMS.[12] The Premium SMS market will likely continue to shrink because, subsequent to the FTC's roundtable, in late 2013, the four largest mobile carriers pledged to discontinue Premium SMS billing for commercial transactions.[13]

In contrast, direct carrier billing, which has generated roughly $300 million in revenue annually, is expected to grow to at least $11 billion worldwide by 2016 for app store purchases alone.[14] DCB arrangements are likely to supplant Premium SMS as the preferred mode of carrier billing, as mobile phone companies have continued to partner with market participants ranging from large technology platforms to smaller mobile payments companies in an effort to make DCB a viable option for mobile payments.[15]

Regardless of the type of carrier billing involved, it is important for companies to keep basic consumer protections in mind, such as providing adequate disclosures containing truthful information and obtaining informed consent for charges. Accordingly, the consumer protection principles embodied in staff's recommendations, discussed in below in Part IV, apply to any Premium SMS arrangements that may still exist or be resumed, DCB arrangements, or other forms of carrier billing that may emerge.

C. Market Participants

The process of placing a charge for third-party services on a phone account typically involves a number of parties, including merchants, billing intermediaries, and mobile carriers. Each of these market participants is under the FTC's jurisdiction when engaged in third-party billing activities.[16]

Merchants provide the goods or services for which consumers are billed on their mobile phone accounts. In Premium SMS arrangements, merchants are often known as "content providers," as many merchants in this area have provided text-based or digital content (such as text-based subscription services or ringtones) to mobile devices. DCB arrangements can be used for many kinds of merchant transactions in which consumers can choose a carrier billing option in lieu of another kind of payment mechanism, like a credit card.

Roundtable participants identified, as examples, merchants that use DCB to sell credits on dating, gaming, or social networking sites.[17]

Merchants sign up with **intermediaries** that have contractual agreements with mobile carriers to place charges on mobile phone accounts. In Premium SMS arrangements, these companies often are known as "aggregators."[18] A range of companies acting as billing intermediaries have begun to move into the DCB space.[19] The consumer may or may not directly interact with the billing intermediary at all.

Mobile carriers bill consumers for these transactions, and the carriers get a portion of the consumer payment. The percentage kept by mobile carriers in third-party billing arrangements varies, though in at least one case (*Wise Media*) the carriers received 30-40% of the amount charged to consumers.[20]

A diagram of the billing process is depicted below:

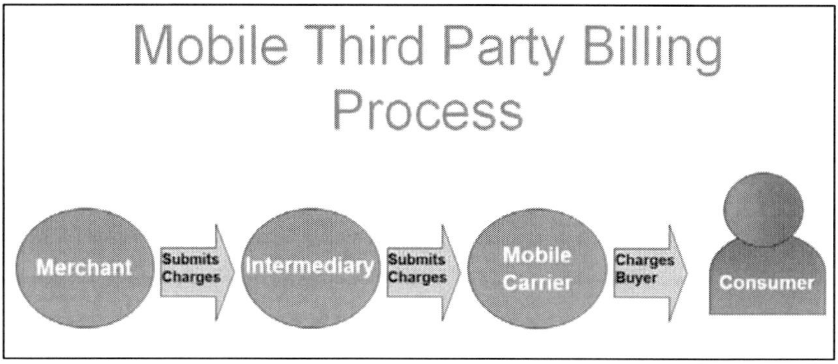

III. THE MOBILE CRAMMING PROBLEM

As the FTC, consumer advocates, and other law enforcement and regulatory authorities have observed, mobile cramming has become a significant problem. Part A of this section discusses what mobile cramming is and provides illustrations of how cramming occurs in the mobile context. Part B discusses the evidence demonstrating the extent of mobile cramming. Part C discusses the experiences of other countries in dealing with cramming.

A. What is Cramming?

Cramming is the placement of an unauthorized third-party charge on a phone account.[21] Cramming on landline phone bills has been a problem for many years – likely costing consumers billions of dollars in the last decade.[22] The FTC has brought over thirty cases to stop landline cramming and return money to consumers.[23] As mobile phones have proliferated in recent years, cramming has emerged as a problem in the mobile arena.

Cramming takes advantage of the fact that many consumers do not know that third-party charges can appear on their phone bills; do not notice the charges, which are often buried in the bill under vague terms such as "usage charges"[24] or "monthly service charges" or other terms that suggest a connection to the carrier;[25] or, in the case of consumers with prepaid plans, do not receive a bill at all. Cramming occurs when consumers are signed up and billed for third-party services by merchants, either without any affirmative action by the consumers or after the consumer takes some affirmative act (such as clicking on a mobile webpage or providing a mobile phone number) without understanding that a charge to a mobile phone account will result. Consumers who receive a monthly bill for their mobile accounts frequently overlook the charges and pay their phone bills in full.[26] Many consumers may also use auto bill-pay or paperless billing, and/or have family plans that cost a hundred or more dollars each month, making them less likely to notice small charges on their bills.[27] And some prepaid mobile phone consumers do not receive a bill at all – crammed charges are simply deducted from their prepaid balance. [28] As many have pointed out, this ability to charge many consumers without their detecting it makes mobile cramming "almost the perfect scam."[29]

1. FTC Actions Addressing Mobile Cramming

Three recent FTC cases in which the FTC reached settlements with content providers alleged to be engaged in cramming illustrate several ways that cramming occurs:

- *Wise Media LLC.* The FTC filed suit in April 2013 against merchant Wise Media, LLC, which purported to sell recurring subscriptions to text message services providing "love tips," horoscopes, diet tips, and similar kinds of "alerts" for $9.99 a month.[30] The company claimed that consumers signed up for the services by entering their information into websites, receiving PIN codes by text messages, and inputting the PINs into the websites. Consumers who discovered the charges, however, widely reported that they had never heard of Wise Media or signed up for the services, suggesting they were simply billed without authorization.[31] In November 2013, a court entered a stipulated order with a judgment for more than $10 million and a ban that prohibits Wise Media from placing charges on mobile phone bills altogether.[32]

- *Jesta Digital, LLC.* In this case, filed in August 2013, the FTC alleged that the defendant lured consumers into purchasing a monthly subscription for ringtones using deceptive practices.[33] According to the complaint allegations, some consumers saw banner ads on their mobile devices that falsely claimed a virus had been detected. Clicking on the ad led to a screen with a button stating "Get Now" above the phrase "Protect your Android [phone] today." Consumers who clicked "Get Now," and then a button on a subsequent page marked "Subscribe," were then subscribed to the $9.99 per month ringtone subscription plan, though the nature and cost of the subscription were never adequately disclosed. Indeed, some consumers were subscribed even if they clicked on parts of the screen other than the "subscribe" button. Moreover, if consumers actually attempted to subscribe and download Jesta's so-called anti-virus software to their mobile devices, the download often failed. Jesta used a process known as WAP or Wireless Access Protocol billing,[34] which captures consumers' phone numbers from a mobile device, to obtain consumers' purported authorization for the charges. Thus, consumers never even entered their phone number prior to being billed.[35]
- *Tatto, Inc. & Bullroarer, Inc.* In a third case, filed in December 2013, the FTC alleged that another widespread mobile cramming operation engaged in similar deceptive practices. For example, the FTC alleged that the defendants ran websites that promised consumers offers such as free Justin Bieber tickets.[36] When consumers attempted to claim these offers, they were asked for a mobile phone number. After following the instructions provided, consumers did not receive the Justin Bieber tickets, yet, the Commission has alleged, it is likely that consumers were instead signed up for the defendants' subscription plans.[37] The primary corporate defendants have agreed to a partially suspended judgment of $150,153,283.[38]

The Commission has also pursued two recent additional actions against content providers that raise similar issues. In one case, the Commission has alleged that certain defendants sent text messages promising free $1,000 gift cards and iPads as a way to deceive consumers into "confirming" their phone number and entering a PIN on a website, which resulted in consumers being signed up for unwanted premium text messaging services and incurring charges of $9.99 per month on their mobile phone accounts.[39] In another case, the Commission has alleged that a content provider similarly used the lure of "free" gift cards to collect consumers' phone numbers and crammed consumers for subscription services such as horoscope alerts.[40]

Additionally, the Commission recently filed suit in federal district court against T-Mobile USA, alleging that T-Mobile charged consumers for these kinds of monthly text message subscriptions purportedly offered by third-party merchants that, in many cases, were not authorized by consumers. The complaint alleges that T-Mobile deceptively described these charges on its phone bills in a manner that made it difficult for consumers to uncover them. For example, for consumers who reviewed an online summary of their bills, the complaint alleges the third-party charges were lumped together in a line item labeled "Use Charges" that could include charges for both T-Mobile's own services, such as for text messages, and for third-party charges. Additionally, according to the complaint, T-Mobile continued to charge consumers even after becoming aware of telltale signs that the charges were unauthorized. The complaint alleges that T-Mobile's internal documents showed that consumers were increasingly calling T-Mobile to complain about unauthorized third-party charges; that large numbers of consumers sought refunds and the refund rate for some subscriptions was higher than 40% in some months; and that T-Mobile continued to charge consumers for third-party merchants for years after those merchants were the subject of law enforcement or other legal action for cramming, news articles detailing cramming behavior, and industry auditor alerts detailing deceptive practices in which those merchants were engaged.[41]

2. Other Federal and State Initiatives Addressing Mobile Cramming

The FTC is joined by a number of partners in its efforts to protect against cramming. On the federal level, for several years the Federal Communications Commission ("FCC") has been engaged in a rulemaking proceeding in which it is proposing additional truth-in-billing rules to help consumers detect crammed charges on phone bills, and is considering whether to extend those rules to mobile bills.[42] The FTC has filed comments in that proceeding calling for stronger consumer protections.[43] And the FCC recently announced it launched an investigation into T-Mobile with regard to cramming.[44]

Congress also has focused on the problem of mobile cramming. The Chairman of the Senate Committee on Commerce, Science, and Transportation has issued requests to mobile carriers and billing aggregators to provide information on cramming in order for the Committee to assess potential legislative solutions.[45] Chairman Rockefeller has noted that, based on review of public complaints by his staff, "[c]onsumers continue to complain that they are experiencing unauthorized charges on their wireless bills for 'services' they did not order and do not use," and "[t]he types of so-called 'services' that are appearing on these consumers' wireless bills . . . are alarmingly similar to the services shown to be fraudulent on wireline telephone bills."[46]

The states have been active in this area as well. Based on their experience reviewing complaints and conducting investigations and enforcement actions, the Attorneys General of 36 States, the District of Columbia, Guam, Puerto Rico, and the Virgin Islands submitted a comment to the FTC, in connection with the roundtable, urging action to address mobile cramming. The comment notes that they "continue to receive complaints from consumers that charges . . . appeared out of the blue on their phone bills without their authorization and for goods and services that the consumers neither requested nor used,"[47] pointing out that the complaints share common themes "consistent across the country and across time."[48] Given the scope of fraudulent conduct, the Attorneys General have noted that they are "concerned that, under this system, too much responsibility has been placed in the hands" of third parties to obtain consumers' consent – "parties for whom there is a significant incentive to bill for services whether or not they have obtained authorization from consumers."[49]

Indeed, multiple state enforcement authorities have been active in combating mobile cramming. States, including Florida,[50] New York,[51] Texas,[52] and Washington,[53] have targeted deceptive practices used to sign up consumers unknowingly for unwanted subscription services. For example, as discussed below in at Part IV.B.1.a, the State of Texas recently sued a mobile cramming operation that allegedly used deceptive means on websites and in text messages to sign up consumers for services without consent, and it has also filed suit against a Premium SMS aggregator, which it alleged participated in merchants' deceptive practices to enroll consumers in the merchants' programs and assisted them in avoiding detection by consumers.[54]

B. Prevalence of Mobile Cramming

One of the key issues discussed by roundtable participants was the prevalence of mobile cramming. Publicly available information, including evidence about the rates at which consumers obtain refunds for mobile third-party charges and the volume of those refunds, indicates that cramming is a significant problem.[55] This section discusses that evidence, as well as indications that these consumer complaints understate the full extent of actual consumer injury.

1. Carrier Refund Rates

Carrier refund rates for commercial third-party billing – the ratio of a carrier's refunds to charges billed for a particular period of time, such as a month – have been significant. Carriers keep track of these refund rates and sometimes keep a greater share of the revenue from third-party charges for merchants that have high refund rates or a high number of consumer complaints than they do for those with lower refund rates or complaints.[56] Between January 2011 and September 2012, mobile carriers reported data to the California Public Utilities Commission (CPUC) showing that the overall refund rate for third-party charges for California mobile consumers was approximately 12-13% each month, with a high of 18.6% in one month.[57] Wise Media, one of the companies the FTC sued for mobile cramming, had refund rates exceeding 40% in some months for certain services.[58]

High refund rates are often considered to be indicia of fraudulent conduct, and the refund rates for commercial carrier billing appear to be an order of magnitude higher than refund rates for other types of billing. For example, charitable donations charged to a mobile bill and processed through the Mobile Giving Foundation typically have a refund rate of under 1% overall.[59] In the credit card industry, public evidence from one payment processing network shows that the average credit card chargeback rate is around 0.2%, and that a chargeback rate of 1% for any one merchant in a month triggers an investigation.[60] Similarly, for ACH (Automated Clearing House) transactions, a 1% rate of chargebacks for unauthorized ACH transactions by a merchant may trigger an enforcement proceeding.[61]

Carriers sometimes set threshold refund rates for disciplining merchants, such as prohibiting them from placing charges on consumers' bills.[62] One carrier, for example, has terminated Premium SMS merchants with an 8% or higher refund rate.[63] Nevertheless, at least one carrier continued billing consumers for Wise Media charges even after the company's monthly refund rate for some services exceeded 40%.[64]

As high as the carrier billing refund rates are, available evidence indicates that they understate the extent of refund requests from consumers. On the one hand, industry representatives stated at the roundtable that carriers have liberal refund policies and that, in some cases, a carrier may provide a refund even if someone else in the complaining consumer's household authorized the charge, or when the complaint is about an issue other than cramming.[65] Consumers, however, have often reported difficulties requesting refunds for crammed charges from carriers. Many complain that carriers refuse to give more than two months' worth of refunds, even if consumers learn that crammed charges have appeared on their bills for longer periods of time.[66] In other instances, carriers have told consumers to contact the merchant for a refund, a request that, even if the consumer is able to make contact with the merchant, the merchant often denies.[67]

2. Complaint Information

Government agencies also have received a significant number of complaints related to third- party charges on mobile accounts. For example, the FTC's Consumer Sentinel database shows that consumers have reported over 1800 complaints of unauthorized charges on wireless bills since 2010.[68] As noted in greater detail below, while this is a significant number, the number of cramming complaints to the FTC likely significantly understates the actual extent of consumer injury. For example, the number is dwarfed by the number of refund requests to carriers.[69] In the *Wise Media* case alone, evidence obtained in the FTC's investigation showed that carriers granted over 190,000 refunds to consumers for Wise Media's charges.[70] And in the *Tatto* case, the FTC's evidence showed that the carriers granted complaining consumers more than 1.2 million refunds for charges by the group of companies the FTC sued.[71] Thus, while the FTC's complaint database showed a total of approximately 1800 complaints, carriers made almost 1.4 million refunds to complaining consumers in two cases alone.

Similarly, the CPUC reports that it has never received more than 10 cramming complaints in a month during the 2011-2012 time period for which it provided data, which is a tiny fraction of the carrier refunds provided during that time. Specifically, for all of 2011 and much of 2012, the CPUC reports that carriers have provided no fewer than 160,000 refunds a month to California consumers and have sometimes provided over 300,000 refunds a month. In every month for which the total number of refunds was reported, carriers provided at least 20,000 times more refunds than complaints received by the CPUC, meaning that complaints reported to the state agency represented less than 0.005% of the refunds actually given in a month.[72]

3. Other Efforts to Estimate the Extent of Cramming

Other stakeholders have also sought to estimate the extent to which third-party charges are crammed as a percentage of the overall carrier billing market. At the roundtable, the Vermont Attorney General's office presented the results of a study addressing the size and nature of the mobile cramming problem in Vermont. The study found that, out of a random sample of Vermont consumers who had incurred third-party charges on their phone bills, 60% of consumers reported that neither they nor anyone in their household had authorized the third-party charges listed on their bills. Over 55% said they were unaware of the charges until the survey.[73] The Citizens Utility Board in Illinois has also announced that an analysis of mobile phone lines by an outside research firm found that 44% of all third-party charges appeared to be fraudulent.[74]

On the other hand, CTIA has criticized the methodology of the Vermont survey and submitted the testimony of an expert who analyzed the Vermont study on behalf of the four major mobile phone carriers. According to CTIA, the expert "found that the survey methodology underlying the Vermont Study did not comply with core principles of objective research, and further concluded that 'the Vermont Study is neither a valid nor reliable measure of the extent to which, if any, Vermont mobile phone users have problems with unauthorized third-party charges on their bills.'"[75]

Although no industry participant attempted to estimate the extent of cramming at the roundtable or in a public comment, industry members also have questioned the prevalence of mobile cramming. CTIA has stated that mobile cramming is "not a significant consumer concern," pointing to the number of mobile cramming complaints reported by federal agencies and the industry's voluntary actions to address cramming.[76] Mobile phone carriers, including Verizon, AT&T, and Sprint, have publicly explained steps they have taken to prevent cramming on the mobile platform and suggest that industry standards are working to prevent fraud in this space.[77] In addition, a representative from the Mobile Giving Foundation, who previously served as chairman of the Mobile Marketing Association, suggested at the roundtable that mobile cramming is on the decline given steps taken by the carriers.[78]

Despite extensive evidence of consumers complaining to and seeking refunds for crammed charges, much of it developed through enforcement actions, industry representatives have not attempted to estimate the full extent of crammed charges based on their own data. Instead, industry representatives primarily rely on data about public complaints to government agencies, which, as noted above, represent a small fraction of actual cramming.

4. Unreported Consumer Injury

The FTC's enforcement actions and other experience further illustrate the fact that reported consumer complaints and refund requests do not reflect the full extent of injured consumers. This is sometimes described as the level of complaints representing the "tip of the iceberg" of actual consumer injury.[79]

- At the time of filing suit in *FTC v. Inc21.com Corp.*, a landline cramming case, the FTC was able to point to about 280 formal consumer complaints to government agencies.[80] An expert survey conducted during the course of litigation and admitted by the court showed that nearly 97% of billed consumers did not authorize the charges.[81] After prevailing in litigation, the FTC provided monetary redress to over 139,357 consumers.[82]
- In *FTC v. Wise Media, LLC*, the FTC noted at the time of filing that it had received over 100 Consumer Sentinel complaints regarding Wise Media's monthly charges of $9.99, but the defendants later agreed to a stipulated judgment amount of over $10 million.[83]
- A 2004 study by the FTC's Bureau of Economics and Bureau of Consumer Protection found that only 8.4% of defrauded consumers complained to a government agency or the Better Business Bureau, meaning that the vast majority of fraudulent conduct is not officially reported to a government agency.[84]

One reason that refund requests and complaints represent only a small fraction of all crammed charges is that many consumers do not notice charges on their phone bills, in part because the descriptions of the charges are often buried and uninformative. As a result, these consumers never submit a complaint or seek a refund at all. In the context of landline cramming, a court-accepted survey in the *Inc21.com* case found that only five percent of consumers were even aware of the unauthorized charges from review of their bills.[85] Several panelists at the roundtable also commented that consumers may not know that their bills can be charged for third-party services, and that many of these small charges can be labeled deceptively.[86] In the *Wise Media* case, for example, consumers complained that the charges were buried in lengthy phone bills (for example, a $9.99 charge was listed on page 18 of one consumer's bill).[87] In another case, the allegedly crammed charge was listed on page 123 of a consumer's online bill.[88] Third-party charges have often been placed in the bill under vague terms such as "usage charges" or "monthly service charges" or other terms that suggest a connection to the carrier.[89] Additionally, some consumers, such as those with prepaid mobile phone plans, may not receive a bill at all.[90]

C. International Views

Consumer protection authorities outside the United States have sought to combat unauthorized third-party charges on consumers' mobile phone bills as well. For instance, Canada's Competition Bureau filed a court action in fall 2012 against three of the major Canadian mobile phone carriers for misleading conduct related to Premium SMS charges.[91] In the United Kingdom, regulators recently have seen a surge in cramming complaints and thus have initiated a series of investigations and have imposed a number of fines.[92] Consumers in other countries, like Australia, Russia, and China, have experienced similar problems.[93] The London Action Plan, an international anti-spam network consisting of 27 countries around the world, released a report in October 2012 called "Best Practices to Address Online and Mobile Threats."[94] In it, they state that "'cramming' to Premium Rate 'love advice' or other text message services by Affiliates and/or Content providers has been commonplace."[95]

Australia has reported some success in combatting mobile cramming. In a case study published by the Australian Competition & Consumer Commission ("ACCC"), the ACCC found that previous controls – including industry self-regulation and some investigation into specific merchants – were not effective in reducing cramming. To address the problem, Australian authorities increased consumer education, brought enforcement actions, and worked closely with industry on adopting a set of regulations to protect consumers, which the ACCC noted was a "crucial component of providing a long-term solution to the problem."[96] Those regulations included, for example, requiring that carriers offer a block on Premium SMS services to consumers.[97] After taking these steps, Australian authorities reported that the number of consumer complaints in Australia about cramming have dropped significantly.[98]

IV. STRATEGIES TO REDUCE MOBILE CRAMMING

Stakeholders in the mobile billing field have adopted a number of strategies to attempt to address unauthorized third-party charges. Below we discuss the current strategies being used to combat mobile cramming in four areas, and provide FTC staff's view on further action to be taken to address the problem. Specifically, industry participants should give consumers the option of blocking third-party charges, implement measures to detect and prevent crammed charges from appearing on mobile phone bills, provide adequate disclosures of third-party charges, and establish clear and consistent dispute resolution policies for unauthorized charges. Staff's recommendations apply regardless of what form of carrier billing is being used, whether it is DCB, Premium SMS, or a new form of carrier billing that emerges.

A. Allowing Consumers to Avoid Third-Party Charges to Mobile Accounts

Consumers should be able to choose to avoid third-party charges to their mobile accounts altogether and, as explained below, some carriers appear to have taken steps to allow consumers to do so. FTC staff reiterates prior Commission recommendations in this area and recommends further industry best practices to empower consumers to make informed choices about third-party charges.

1. Current Industry Practices

According to CTIA, the four major carriers currently permit their customers to block third-party charges from being placed on their phone accounts.[99] This option gives consumers who may not wish to use their phone accounts as a payment mechanism the ability to avoid third-party charges altogether, including unauthorized charges.[100] Although some carriers provide information about these blocking options in their service agreements, at the point of sale, and on their consumer-facing websites, several stakeholders have suggested that customer service representatives are poorly informed about the option, and that mechanisms to add or remove a block have not always been prominently featured or clearly described in the service agreement or on carrier websites.[101] Thus, some consumer groups have recommended prohibiting all third-party charges on wireless accounts by default, with an exemption for charitable giving.[102]

During the mobile cramming roundtable, industry participants stated that some third-party charges originate on devices used by children.[103] When children incur charges on a family plan, the primary account holder – typically a parent or other adult – receives a single monthly bill that includes charges for all the family members' mobile devices that share the account. According to CTIA, mobile carriers offer account holders the opportunity to block all third-party charges on a family account or to block individual mobile devices from adding charges to the account.[104] As discussed above in Part III, however, evidence shows that many consumers do not understand that mobile telephone bills may contain third-party charges.

2. FTC Staff Recommendations

As the Commission stated in comments to the FCC, all wireless providers should give consumers the option to block all third-party charges from their mobile phone accounts.[105] Providing a blocking option would significantly benefit consumers who wish to avoid third-party charges while imposing minimal costs to consumers who wish to use their mobile accounts for third-party billing.[106] At activation, consumers should be informed that third-party charges may be placed on their accounts, including using any phone numbers on a family plan, and they should be given the opportunity to block all charges at that time. This option should be clearly and prominently disclosed to consumers while the accounts are active, including on the carriers' websites. According to CTIA, carriers currently offer the ability to block third-party charges on specific phone numbers on a family plan.[107] Given the concerns raised about cramming on children's numbers, carriers should clearly and prominently disclose these options as well. Notably, the FCC's current Truth-in-Billing rules require landline carriers that offer blocking of third-party charges to clearly and conspicuously notify consumers of this option on each bill, their websites, and at the point of sale.[108]

Additionally, carriers should consider offering consumers the ability to block or allow only specific providers, or to block commercial providers only. In the context of Premium SMS, some carriers provided the option to block individual short codes or subscriptions.[109] Allowing more granular blocking would allow consumers to continue to authorize some third-party charges, including charitable or political donations, and may significantly benefit consumers who wish to use their mobile accounts for only certain kinds of third-party charges.[110]

B. Strategies to Detect and Prevent Mobile Cramming

Industry participants have adopted a range of strategies to attempt to detect and prevent mobile cramming. This section focuses on strategies involving two key issues: avoiding deceptive practices that lead to unauthorized charges on mobile accounts, and ensuring that consumers are providing express, informed consent to third-party charges on mobile accounts. As explained below, industry efforts have fallen short on both fronts. For example, carriers are not taking sufficient action against merchants with high refund rates. Further, the voluntary standards developed by industry have largely focused on text-message based Premium SMS services and have not specifically addressed other types of carrier billing including app- or mobile web-based billing using DCB arrangements. Accordingly, staff recommends best practices for improvement in this area.

1. Avoiding Deceptive Practices That Lead to Unauthorized Mobile Charges

a. Industry Practices

Stakeholders in the mobile billing industry generally have relied on a set of voluntary guidelines to prevent deceptive marketing or advertising of mobile-billed goods or services. As discussed below, various industry participants monitor merchants' marketing and advertising practices, and track merchants who have previously engaged in deceptive behavior, to varying degrees. Despite these industry best practices, however, there have been significant problems with enforcement and compliance. Moreover, it is unclear how these best practices – many of which were developed in connection with the text-message based Premium SMS model – are being applied in the context of DCB arrangements.

Voluntary guidelines. Until recently, the Mobile Marketing Association ("MMA"), a trade association that promotes mobile marketing, had taken the lead in publishing best practices for merchants who wish to place charges on mobile phone bills using Premium SMS.[111] In January 2013, CTIA began publishing a "Mobile Commerce Compliance Handbook," which aims to distill the principles of the MMA best practices to a shorter set of guidelines. The CTIA guidebook which describes itself as a "unified standard of compliance for mobile carrier billing," notes that its "requirements are based on the CTIA and participating carriers' experience with standard rate and premium shortcode [Premium SMS] programs," but that "the core rules are relevant to many types of mobile services."[112] It is not clear, however, how or to what extent the CTIA guidelines are used across all carrier billing arrangements.

The MMA's best practices include fairly detailed requirements for merchants to advertise their products and obtain consumers' authorization for third-party charges billed to a mobile phone account. Some commenters have noted, however, that merchants technically could comply with guidelines while hiding key information, such as the fact that entry of a mobile phone number to access purportedly free content will automatically subject the consumer to a recurring $9.99 charge on a monthly phone bill.[113] Specifically, a set of related companies that the State of Texas sued for cramming technically complied with an MMA guideline that a price disclosure be 125 pixels away from a box to enter a cell phone number on a website, but blended the price disclosure with a background image to make it difficult to see, as illustrated on the following page:[114]

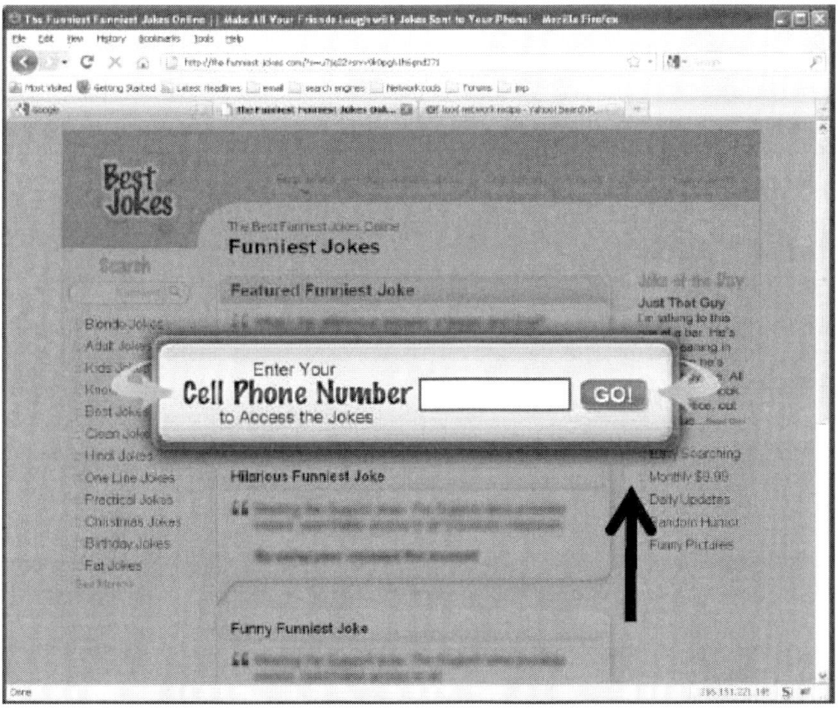

(vertical arrow added.) The defendants sent confirmation texts to consumers that included large gaps of spacing in the text, so that the consumer was likely to see the PIN code for opt-in at the top or bottom of the message, but unlikely to see the price disclosure buried inconspicuously in the middle of the text message, which would be unlikely to be displayed when the message was opened, as shown below:[115]

Bottom of Message (shown when opened)

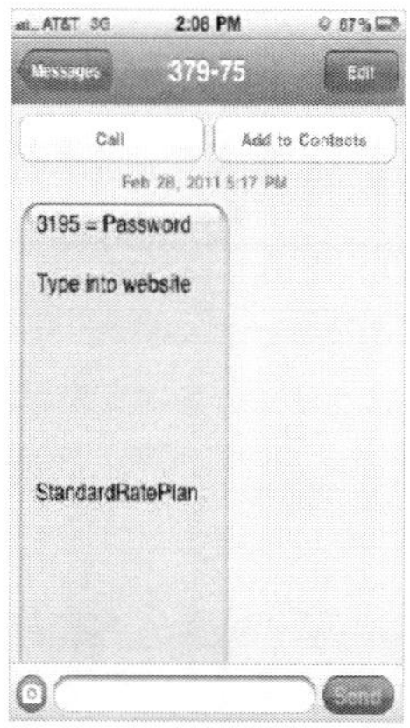

Top of Message

Middle of Message

Monitoring and Enforcement. Although the MMA first assembled the set of best practices that became known as "guidelines" for the industry, it has no role in monitoring industry participants to enforce the guidelines.[116] Instead, industry participants choose whether to monitor and take action against merchants that violate the guidelines or otherwise engage in deceptive advertising or marketing practices.[117]

For example, carriers decide whether to allow merchants to bill on their networks, and prior to such approval, most carriers will review merchants' marketing materials and opt-in processes to ensure that they comply with the industry guidelines.[118] Each carrier then determines how much effort to invest in monitoring for compliance. At least some carriers have performed "in-market monitoring" – often by hiring third-party auditors – to check that the advertising and purchase flow continue to be compliant with industry or carrier-specific guidelines, at times looking beyond mere compliance with

MMA and CTIA guidelines.[119] Further, at least one aggregator has noted that it also has engaged in in- market monitoring of campaigns.[120]

To be reasonably effective, this kind of monitoring must take into account challenges from determined fraudsters. For example, one entity used sophisticated cloaking software that displayed non- deceptive webpages to auditing companies, but displayed a different webpage to consumers.[121] Industry actors state that they have learned from experience in dealing with this kind of evasive activity.[122] Beyond reviewing and monitoring merchants' marketing practices for deceptive behavior, CTIA and some carriers also have vetted merchants up front based on their past practices – in short, attempting to identify potential bad actors in advance and prevent them from billing on carrier networks. While this monitoring of relationships can identify individuals or entities associated with fraudulent behavior in some contexts, it does not necessarily look closely at whether the merchant has been involved with unauthorized billing on landline phone bills, which some commenters suggest would be appropriate given the prevalence of cramming on landline phone bills.[123]

b. Staff Recommendations

Under the FTC Act, merchants are in the first instance responsible for ensuring that their practices – including any advertising, marketing, and opt-in processes – are not deceptive, pursuant to the FTC Act. Further, information about price is important to consumers and should be disclosed clearly and conspicuously before charging a consumer's telephone account for a good or service. As FTC staff stated in its 2013 report, *.com Disclosures: How to Make Effective Disclosures in Digital Advertising*,

> Disclosures that are an integral part of a claim or inseparable from it should not be communicated through a hyperlink. Instead, they should be placed on the same page and immediately next to the claim, and be sufficiently prominent so that the claim and the disclosure are read at the same time, without referring the consumer somewhere else to obtain this important information. This is particularly true for cost information or certain health and safety disclosures.[124]

Thus, at a minimum, pricing information should be on the same page and immediately next to the purchase or buy button, entry of a PIN, or other invitation for a consumer to agree to a charge for a product or service, whether through a Premium SMS, DCB, or some other arrangement, and this price information should be prominent and in a font and size sufficient to read.

Further, advertising and purchase confirmation screens must clearly disclose that the charge is being billed to a specific telephone account. Companies may consider, for example, whether identifying the last four digits of the consumer's phone number on a purchase confirmation screen – similar to the way that the last four digits of a credit card number are commonly listed on confirmation screens – is feasible and useful as part of the key disclosure that the charge will be placed on a telephone account.[125] While industry guidelines have in the past focused extensively on the text-message based Premium SMS opt-in process, the basic consumer protection principles outlined here should apply regardless of the type of carrier billing used, whether Premium SMS or DCB.

Staff also recommends that carriers and billing intermediaries implement reasonable procedures to scrutinize risky or suspicious merchants and terminate or take other appropriate steps against companies engaging in unlawful practices. These kinds of measures are present in other payment industries. In the credit card and ACH industries, for example, banks that provide payment processing services are subject to KYC ("Know Your Customer") procedures, which require banks to perform due diligence on their merchant-customers. Among other things, KYC principles require banks to gain familiarity with a merchant's marketing practices to identify high-risk activities and ensure that it is not facilitating fraudulent or other illegal activity.[126] Further, the operating rules established by the credit card associations and NACHA – The Electronic Payments Association (for ACH transactions) encourage banks to monitor merchant transaction activity to identify merchants engaged in fraud or other illegal conduct, and to cut off their access to the payment systems if appropriate.[127] While the specific procedures may vary by industry, this kind of monitoring and termination of fraudulent merchants has been beneficial in protecting consumers and maintaining the integrity of the billing platform.

Similarly, staff recommends that if a carrier or billing intermediary discovers that a merchant has run a campaign containing deceptive advertising, or discovers the merchant engaged in unauthorized billing on landline phones, the carrier should closely monitor other campaigns run by that third party or its affiliates to ensure compliance. Carriers can use monitoring techniques to address known tactics by fraudsters to evade detection, in order to determine consumers' actual experience with third-party merchants' advertisements and sign-up processes, such as viewing advertisements on devices not registered to the carrier. Industry participants also can adopt a policy of terminating serious and repeat offenders. While there are costs to effective monitoring, there are also benefits both to the industry itself, in the form of lower costs to process refund requests and handle customer complaints, and to consumers, who avoid being crammed with unauthorized charges.

2. Obtaining Consumers' Express, Informed Consent

a. Current Industry Practices

As discussed above, mobile industry participants created the MMA best practices and the CTIA guidelines in an effort to ensure that consumers authorize the third-party charges that mobile carriers include on mobile telephone accounts. In the context of Premium SMS, the core of those self-regulatory initiatives is a requirement that third-party merchants employ a "double opt-in" process that requires consumers to take two separate steps to authorize a third-party charge, one of which is intended to confirm that the consumer is in physical possession of a phone tied to the account that will be charged.[128] As noted above, it is not clear whether industry participants in DCB arrangements routinely follow the same sort of double opt-in process used for Premium SMS transactions or use other methods to obtain consumers' authorization.

Individual carriers have also used refund rates to identify third-party merchants who may not be obtaining consumer consent to charges, and individual carriers have the discretion to penalize or cut off merchants with high refund rates,[129] with one carrier, for example, terminating Premium SMS merchants with an 8% or higher refund rate.[130] Based on the record of consumer complaints, refunds, and law enforcement actions described above, however, it does not appear that the double opt-in requirement or carriers' current practices involving monitoring refund rates and terminating problematic merchants have prevented unauthorized charges on consumers' mobile telephone accounts. Commenters have noted that crammers can easily obtain consumers' phone numbers and find ways to bill consumers even without completing either step of the purported double opt-in.[131] Consumers have repeatedly complained that charges appear on accounts without *any* affirmative action by the consumer.[132]

Given the issues with spurious consumer authorizations, one question raised by various stakeholders is which entity in the carrier billing process should host the records and control the process of the consumer authorization. Although merchants historically have controlled the opt-in process for Premium SMS,[133] there appears to be a trend toward greater carrier or aggregator control of the opt-in process.[134] With respect to noncommercial third-party charges, one representative of the Mobile Giving Foundation, a billing intermediary, noted that it manages the text-based opt-ins for participating charities.[135] And at least some DCB intermediaries also appear to manage the consumer opt-in, including by sending the consumer a PIN to confirm authorization.[136] Centralization allows for greater control and transparency as to individual merchants' authorization and billing activities.[137] One direct carrier billing intermediary has stated that managing the two-step confirmation process is one of the factors that has led to its low refund rate, reportedly less than 1% on some carriers.[138]

As mobile cramming issues have become more apparent, state regulators and other stakeholders have called for greater centralization of opt-in records as one way to address cramming.[139] The Attorneys General from thirty-six states, the District of Columbia, Guam, Puerto Rico, and the Virgin Islands have criticized current industry practices for placing too much of the opt-in responsibility in the hands of third-party merchants and billing intermediaries. In the view of those Attorneys General, those parties have significant incentive to bill for services whether or not consumers have authorized them.[140]

An order by the CPUC in 2010 sets forth the general principle that "the billing telephone corporation [is the entity that] bears ultimate responsibility for all items presented in a subscriber's bill."[141] In 2013, the National Association of State Utility Consumer Advocates (NASUCA) recommended the passage of state legislation to address mobile cramming.[142] The proposed legislation would hold telephone companies, aggregators, and third-party merchants all responsible for unauthorized charges, providing a defense only if "the circumstances giving rise to the unauthorized charge were beyond the control of, or could not reasonably have been prevented by, the company."[143]

b. Staff Recommendations

It is critical that consumers provide their express, informed consent to charges before they are billed to a mobile account, and that reliable records of such authorizations are maintained. Industry practices have not sufficiently prevented unauthorized charges. In Premium SMS, mobile carriers typically have relied on the merchant's representation – passed on by the billing intermediary – that a consumer opted-in to a charge, but, as the law enforcement actions and consumer complaints described above demonstrate, those representations often are unreliable.

Given the unreliability of many merchants' claims that they have obtained consumer consent, more centralized control of the consumer opt-in process and authorization records is needed. As noted above, the industry appears to have already implemented changes in this direction. As the technology used for consent management shifts from Premium SMS arrangements to DCB arrangements, carriers and intermediaries should maintain sufficient control over the consent process to address the problem of unauthorized charges, for example, by maintaining reliable and accessible records of consumer authorizations. Centralization may shift some compliance costs from the merchants to carriers and intermediaries, but should benefit consumers and industry participants by making it more difficult for unscrupulous merchants to place unauthorized charges and by streamlining dispute resolution when a consumer claims a charge was unauthorized.

Further, individual carriers' policies on taking action against merchants with high refund rates have not been sufficient to combat mobile cramming – for example, as noted above, at least one carrier continued billing consumers for Wise Media charges even after the company's monthly refund rate exceeded 40% on some campaigns.[144] Mobile carriers should implement policies, or strengthen existing polices, to investigate and take appropriate action when consumer refund requests and complaints indicate that a merchant may be cramming charges without consumers' consent. As discussed above, other industries have examined refund rates to determine what rates should trigger additional scrutiny and possible termination of a merchant.[145] Monitoring consumer refund requests, and taking appropriate action when there are indications of unauthorized charges, can be a highly effective means of detecting and stopping cramming. It is particularly important to monitor refund requests in carrier billing, where merchants have been able to cram charges without using deceptive advertisements or other techniques to contact consumers that can be independently identified.

C. Adequate Disclosures of Third-Party Charges

Another important step in preventing cramming is ensuring that consumers are adequately informed of all third-party charges on their accounts. Many consumers currently overlook third-party charges on their mobile accounts. Staff provides recommendations in this section for making disclosures of those charges clearer and more prominent.

1. Current Industry Practices

As discussed above in Part III, many consumers have found it difficult to detect unauthorized charges on their mobile phone bills because of the description and location of third-party charges on the bill.[146] For example, the State of Texas alleged in one case that crammed charges were sometimes listed as "Standard Rate Plan" on the consumers' bills.[147] In the FTC's case against Wise Media, the charges typically appeared buried among other charges; as noted above, one consumer reported that a Wise Media charge was listed on page 18 of her phone bill. Wise Media charges also appeared in a manner that made it difficult, if not impossible, to identify that the charges were for services provided by a company other than the carrier.[148] And in its complaint against T-Mobile, the Commission has alleged that consumers who reviewed an online summary of their bills would see third-party charges lumped together in a line item labeled "Use Charges" that could include charges for both T-Mobile's own services, such as for text messages, and for third-party charges.[149] Roundtable panelists noted that many consumers are simply unaware that third parties can even place charges on mobile phone bills, and thus are unlikely to review complicated and lengthy phone bills to find such charges and make sure they are authorized.[150]

The FCC's current Truth-in-Billing rules require mobile carriers' bills to clearly identify the service provider and provide a full and non-misleading description of each charge.[151] In the landline space, there are additional FCC requirements that carriers place all third-party charges for non-telecommunications services in a distinct bill section separate from all carrier charges, and that the bill must provide separate totals for carrier and non-carrier charges.[152] Although these latter requirements currently apply to landline bills only, the FCC recently reopened its request for comment on whether to expand them to mobile phone bills.[153]

2. Staff Recommendations

To help consumers understand what third-party services they are paying for with their mobile phone accounts – and determine whether there are any unauthorized charges on their accounts – carriers should clearly and conspicuously disclose all charges for third-party services in a non-deceptive manner. In particular, the name of the third-party service and any associated bill heading should relate to the product offered and not suggest an affiliation with the carrier's service. Billing intermediaries and merchants must provide accurate information to carriers for these disclosures.

FTC staff further recommends that carriers, as well as the FCC, consider ways to make third- party charges more conspicuous. For example consistent with the FCC's rules for landline billing, third- party charges could be placed in a separate part of consumers' mobile phone bills and separate subtotals for carrier and third-party charges be provided wherever total charges are disclosed – typically on the first page of the bill. As discussed above, many consumers are unaware that third-party charges can be placed on their mobile phone bills, and consumers have expressed frustration that the charges are often buried in lengthy phone bills. The benefits of enhanced disclosures, which will help consumers identify unauthorized charges, would likely outweigh the costs of any necessary changes to the format of the mobile phone bill.

Staff notes that these recommendations may be less beneficial for consumers who auto-pay their bills and may be especially unlikely to review the charges. Carriers should thus consider whether these consumers would benefit from receiving a separate notification of third-party charges. Consumers with prepaid phone plans who do not typically receive phone bills should also receive such a notification from the carrier.

D. Consumer Dispute Protections and Refunds

Stakeholders have expressed widely divergent views regarding consumers' ability to obtain refunds for unauthorized third-party charges under current industry practices. As noted below, industry representatives state that the carriers have generous refund policies that enable consumers to obtain refunds promptly from their carriers when they believe they have been crammed.

By contrast, consumer advocates and regulators state that it is difficult for consumers to obtain refunds, and they also note that refunds often are limited to one or two months' worth of charges even when consumers discover they incurred crammed charges for a longer time period. FTC staff reiterates the Commission's prior recommendation that mobile carriers should provide a clear and consistent process for customers to dispute suspicious charges on their mobile accounts and obtain reimbursement.[154] Staff provides further recommendations for industry in implementing that process.

1. Current Industry Practices

Two major carriers submitted comments stating that they encourage consumers to contact them to discuss third-party charges or billing issues.[155] Further, according to CTIA, mobile carriers have generous refund policies with respect to third-party charges on consumer's telephone bills.[156] In particular, CTIA touts carriers' "one-and-done" policies, which empower the carriers' customer service representatives to address a consumer's complaint about unauthorized third-party charges during the consumer's initial call to dispute the charge.[157] Additionally, in the landline context, the United States Telecom Association has stated that non-payment of third-party charges will not result in phone service being suspended or terminated.[158] Carriers, however, have not stated that this protection applies to the mobile context.

Consumers' experiences attempting to get refunds vary and are not consistent with a "one-and- done" policy.[159] In lawsuits that alleged cramming, many consumers reported that they found it difficult, if not impossible, to obtain refunds from their carriers when they disputed third-party charges.[160] Some consumers reported that they were only able to obtain partial refunds rather than full refunds for the entire period during which they incurred unauthorized charges.[161] Law enforcement officials have also noted that consumers have inconsistent experiences with obtaining refunds when they discover that they have been crammed.[162] Consumers' experiences include carriers providing partial refunds, carriers providing no refunds and instead referring consumers to the third-party merchant, and third-party merchants providing partial or no refunds.[163] In addition to calling for clear and consistent dispute policies, consumer advocates also recommend that consumers be able to withhold payment on disputed charges with the knowledge that their phone service will not be cut off while the investigation is pending.[164] Although the landline third-party billing industry states that it has taken this step, one roundtable participant raised concerns that consumers are unaware that they can withhold payment pending a dispute of a third-party charge on a mobile account.[165]

It is also not clear whether, and to what extent, carriers proactively notify or provide refunds to consumers who have been billed by merchants that the carrier has terminated for engaging in cramming. According to CTIA, "in appropriate cases," at least some carriers will proactively provide retroactive refunds without waiting for consumers to complain.[166] The carriers have not, however, provided any data regarding the frequency with which they have issued retroactive refunds to all consumers who incurred charges from a third party that the carriers terminated for cramming.

2. Staff Recommendations

As the Commission has previously emphasized, carriers should implement a clear and consistent process for consumers to dispute suspicious charges on their mobile accounts and obtain refunds for unauthorized charges.[167] In the FTC Mobile Payments Report, FTC staff noted that no federal statutory protections have been applied to consumer disputes about unauthorized charges placed on mobile carrier accounts, in contrast to the implementation of statutory dispute rights for unauthorized credit card and debit card charges.[168] For example, in the credit card industry, consumers have dispute resolution rights and liability limits for unauthorized charges under Regulation Z, including a right to withhold payment while the dispute is pending.[169] For debit card users, Regulation E provides protection for consumers, including a requirement that funds debited in an unauthorized transaction be returned to a consumer's account within ten days, pending further investigation.[170]

Because consumers currently do not have access to a clear and consistent process for disputing unauthorized charges to their mobile accounts, staff recommends that carriers implement such processes and ensure that their customer service representatives abide by them. For example, as noted above, in the landline context, industry members have stated that consumers can withhold payment on disputed charges during the dispute period without a cut-off in phone service or accrual of interest. Consumers appear to be unaware of an analogous right in the mobile billing context, however. Staff recommends that this protection be extended to the mobile billing context, and that consumers be informed of it. Further, given the extensive evidence that consumers are unaware of third-party charges on their phone accounts, FTC staff recommends that when consumers seek refunds for recurring unauthorized charges and a carrier concludes those charges were crammed, consumers could be granted refunds for the same recurring charge in previous months to the extent it is practicable to identify those prior charges.

Finally, when a third party's billing activities are terminated for unauthorized charges, staff recommends that the carrier notify consumers who incurred charges from the third party to allow them to request a refund.[171] Moreover, given the major carriers' recent decision to discontinue Premium SMS services in response to documented allegations of fraud,[172] carriers should consider proactively notifying consumers who have paid a Premium SMS charge for merchants that the carriers have reason to believe are crammers and informing those consumers of the steps they should take to seek a refund if they believe the charge was unauthorized. In these circumstances involving clear fraud, consumers should have an opportunity to be made whole.

CONCLUSION

As consumers increasingly turn to their mobile phones as a payment mechanism, it is critical that carriers and other industry participants proactively address mobile cramming. Although the industry has taken some steps to combat unauthorized third-party charges on mobile phone accounts, more needs to be done to address this ongoing problem. Companies must keep basic consumer protection principles in mind as they develop and promote carrier billing options, particularly as the industry continues its transition from Premium SMS to other billing platforms like DCB. Mobile billing offers a promising option for consumers, but industry participants must recognize that innovation goes hand-in-hand with longstanding consumer protection principles. The more companies keep these principles in mind, the more consumers will trust and adopt these innovative payment methods.

The FTC will continue to monitor and, where appropriate, investigate industry participants – carriers, billing intermediaries, and merchants – involved in third-party mobile billing and bring further enforcement actions. Further, the FTC will continue to monitor the issue of cramming on mobile phone accounts and evaluate whether other potential solutions – including legislative measures and additional regulatory changes – are necessary to ensure consumers are protected from unwanted and unauthorized charges.

End Notes

[1] *See* Cecilia Kang, *A Nation Outnumbered By Gadgets*, WASHINGTON POST, Oct. 12, 2011 *available at* http://www.washingtonpost.com/business/economy/a-nation-outnumbered-by-gadgets/2011/10/11/gIQAhjdhdL_story.html; BD. OF GOVERNORS OF THE FED. RESERVE, CONSUMER AND MOBILE FINANCIAL SERVICES 2014, 4-5 (2014), *available at* http://www.federalreserve.gov/econresdata/consumers-and-mobile-financial-services-report-201403.pdf [hereinafter "MOBILE FINANCIAL SERVICES REPORT"].

[2] *See* Cary Stemle, *Direct Carrier Billing: The world's most popular mobile payment*, MOBILE PAYMENTS TODAY (Oct. 15, 2013) http://www.mobilepaymentstoday.com/ blog/11377/ Direct-Carrier-Billing-The-world-s-most- popular-mobile-payment-Infographic.

[3] *FTC v. T-Mobile USA, Inc.*, No. 2:14-cv-00967-JLR (W.D. Wash.); *FTC v. Wise Media, LLC*, No. 1:13-cv-1234-WSD (N.D. Ga.); *FTC v. Jesta Digital, LLC*, No. 1:13-cv-01272-JDB (D.D.C.); *FTC v. Tatto, Inc.*, No. 2:13-cv-08912-DSF-FFM (C.D. Cal.); *FTC v. Acquinity*

Interactive, LLC, No. 0:14-cv-60166-RNS (S.D. Fla.) (amended complaint filed June 16, 2014); *FTC v. MDK Media, Inc.*, No. 2:14-cv-05099-JFW-SH (C.D. Cal.).

[4] *See* Reply Comment of the Fed. Trade Comm'n, FCC CG Docket No. 11-116 (July 20, 2012), at 7, 12, *available at* http://www.ftc.gov/sites/default/files/documents/advocacy documents/ftc-reply-comment-federal- communications-commission-concerning-placement -unauthorized-charges/120723crammingcomment.pdf [hereinafter "FTC Reply Comment"].

[5] *See* Press Release, Fed. Trade Comm'n, FTC to Host Mobile Cramming Roundtable May 8 (Mar. 8, 2013), *available at* http://www.ftc.gov/news-events/press-releases/2013/03/ftc-host-mobile-cramming-roundtable-may-8.

[6] *See* FED. TRADE COMM'N STAFF, PAPER, PLASTIC... OR MOBILE? AN FTC WORKSHOP ON MOBILE PAYMENTS 7-8 (2013), *available at* http://www.ftc.gov/sites/default/files/ documents/reports/paper-plastic-or-mobile-ftc- workshop-mobile-payments/p0124908 _mobile_payments_workshop_report_02-28-13.pdf [hereinafter "FTC MOBILE PAYMENTS REPORT"]. *See also* Sprint Nextel, *Comments for the FTC Mobile Cramming Roundtable* (May 6, 2013), at 1, *available at* http://www.ftc.gov/sites/default/files/documents/public _comments/2013/05/564482-00008-85922.pdf, [hereinafter "Sprint Nextel Comments"]; Verizon Wireless, *Written Comments for the FTC Mobile Cramming Roundtable* (May 8, 2013), at i, *available at* http://www.ftc.gov/sites/default/files/ documents/ public_ comments/2013/05/564482-00010-85928.pdf, [hereinafter "Verizon Wireless Comments"]; *Manage Mobile Purchases and Subscriptions*, AT&T, http://www.att.com/esupport/ article.jsp?sid=52709&cv=820&title=AT#fbid=1T6cx3Wnq50 (last visited July 23, 2014). "Mobile billing" and "carrier billing" refer herein to the general practice of charging a consumer's phone account for third-party services, whether or not the consumer receives a bill. As noted at III.A, *infra*, many consumers, such as consumers with prepaid accounts, do not actually receive a bill.

[7] *See* MOBILE FINANCIAL SERVICES REPORT, *supra* note 1, at 1-2, 5. Beyond the United States, carrier billing may be the only available electronic payment channel for 1.7 billion people in the world who own a mobile phone but do not have a bank account. *See, e.g.*, Oded Israeli, *The "3x factor" of carrier billing in app store purchases*, MOBILE PAYMENTS TODAY (Feb. 1, 2013) http://www.mobilepaymentstoday.com/blog/9789/The-3X-factor-of-carrier-billing-in-app-store-purchases.

[8] *See* Transcript of the Fed. Trade Comm'n Mobile Cramming Roundtable (May 8, 2013), J. Greenwell, BilltoMobile, at 17, *available at* www.ftc.gov/sites/default/files/documents/ public_events/Mobile%20Cramming%20Roundtable/30508mob.pdf [hereinafter "FTC Roundtable transcript"]; *see also id.*, M. Niejadlik, Boku, Inc., at 135. References to the FTC Roundtable transcript identify the speaker, the transcript page, and, at the first reference to a particular speaker, the speaker's place of employment.

[9] *See* FTC Roundtable transcript, J. Manis, Mobile Giving Found., at 15-16; *id.*, J. Breyault, National Consumers League, at 31; *id.*, D. Asheim, Give by Cell, at 131-32. *See also* AARON SMITH, PEW INTERNET PROJECT, REAL TIME CHARITABLE GIVING: WHY MOBILE PHONE USERS TEXTED MILLIONS OF DOLLARS IN AID TO HAITI EARTHQUAKE RELIEF AND HOW THEY GOT THEIR FRIENDS TO DO THE SAME (Jan. 12, 2012), *available at* http://www.pewinternet.org/~/media//Files/Reports/2012/Real%20Time%20Charitable%20 Giving.pdf.

[10] *See* FTC Roundtable transcript, A. Sege, mQube, at 142. *See also* Derek Johnson, *Political Campaign Text Message Donations – What You Need To Know*, TATANGO (June 11, 2012), http://www.tatango.com/blog/political-campaign-text-message-donations-what-you-need-to-know/.

[11] *See* FTC Roundtable transcript, J. Greenwell, at 23; *see also id.*, J. Breyault, at 14.

[12] *See* FTC Roundtable transcript, J. Manis, at 21; *id.*, J. Greenwell, at 23, 64; *id.*, K. McCabe, Office of Vermont Attorney General, at 54.

[13] Ina Fried, *AT&T, Sprint, T-Mobile, Verizon Dropping Most Premium Text Service Billing in Effort to Combat Fraud,* ALL THINGSD.COM (Nov. 21, 2013), http://allthingsd.com/

20131121/ att-sprint-t-mobile-verizon-all- dropping-most-premium-text-service-billing-in-effort-to-combat-fraud/.

[14] *See* FTC Roundtable transcript, J. Greenwell, at 17, 21, 64; *id.*, M. Niejadlik, at 184; Nick Holland & Rich Karpinski, *Carrier Billing: The Latent Operator Opportunity*, YANKEE GROUP (Aug. 21, 2012), *available at* http://www.yankeegroup.com/Research Document.do?id=59249; JORDAN MCKEE, YANKEE GROUP, THE FALL OF PREMIUM SMS PAVES THE WAY FOR DIRECT OPERATOR BILLING DOMINANCE (Feb. 2014), http://bango.com/_/data/support/Fall_of_Premium_SMS.pdf?b_c=YG&b_s=1 &b_ct=img.

[15] *See, e.g.*, FTC Roundtable transcript, J. Greenwell, at 17, 21, 64; *id.*, M. Niejadlik, at 135, 184; Jessie Xu, FACEBOOK, *Helping Monetize the Mobile Web* (June 6, 2012), *available at* https://developers.facebook.com/blog/post/2012/06/06/helping-monetize-the-mobile-web/; *Mobile Operator Billing*, MICROSOFT, http://msdn.microsoft.com/en- us/library/ windowsphone/help/jj215902%28v=vs.105%29.aspx (last visited July 23, 2014).

[16] *See* FTC MOBILE PAYMENTS REPORT, *supra* note 6 at 2-3; *FTC v. Verity Int'l, Ltd.*, 443 F.3d 48, 59-60 (2d Cir. 2006); *In re Detariffing of Billing and Collection Servs.*, 102 F.C.C.2d 1150 30-34 (1986).

[17] *See* FTC Roundtable transcript, J. Greenwell, at 17, 20-21; *id.*, M. Niejadlik, at 135-36.

[18] *All Aggregators*, COMMON SHORT CODE ADMINISTRATION, http://www.usshortcodes.com/ partners/ find-a-sms- marketing-partner.php#aggregators-tab (last visited July 23, 2014).

[19] *See* FTC Roundtable transcript, J. Greenwell, at 17; *id.*, M. Niejadlik, at 135-136; *id.*, A. Sege, at 143. Companies such as Facebook and Skype have partnered with intermediaries to enable carrier billing in some circumstances on their platforms. *See* Ingrid Lunden, *Facebook Mobile Payment Via Carrier Billing (and Bango) Live In U.S., UK and Germany*, TECHCRUNCH (Sept. 24, 2012), http://techcrunch.com/2012/09/24/facebook- mobile-payments-via-carrier-billing-and-bango-now-live-in-u-s-uk-and-germany/; Ingrid Lunden, *Skype Gets Closer to Mobile Carriers, Inks Deal With Mach For Direct Billing For Skype Credits,* TECHCRUNCH (Sept. 5, 2012), http://techcrunch.com/2012/09/05/skype-gets-closer-to-carriers-inks-deal-with-mach-for-direct-billing/.

[20] *See* Declaration of Andrew R. Schlossberg (Dkt. #3-1), *FTC v. Wise Media, LLC*, No. 1:13-cv-01234-WSD (N.D. Ga. Apr. 16, 2013), at FTC 1 - 000017, 38 [hereinafter "Wise Media Declaration"]; *see also* Complaint for Permanent Injunction and Other Equitable Relief, *FTC v. T-Mobile USA, Inc.*, No. 2:14-cv-00967 (W.D. Wash. July 1, 2014), 10 [hereinafter "T-Mobile Complaint"] (alleging that T-Mobile retained typically at least 35% of the charge and in some cases as high as 40%).

[21] This report does not address unauthorized carrier charges on mobile phone bills.

[22] MAJORITY STAFF OF S. COMM. ON COMMERCE, SCI., & TRANSP., OFFICE OF OVERSIGHT & INVESTIGATIONS, UNAUTHORIZED CHARGES ON TELEPHONE BILLS (July 12, 2011), at ii, *available at* http://www.commerce.senate.gov/public/?a=Files.Serve&File_id=3295866e-d4ba-4297-bd26-571665f40756 [hereinafter "S. COMMERCE COMM. CRAMMING REPORT"].

[23] *See, e.g.*, *FTC v. Inc21.com Corp.*, 745 F. Supp. 2d 975 (N.D. Cal. 2010), *aff'd*, 2012 WL 1065543 (9th Cir. Mar. 30, 2012); *FTC v. Hold Billing Servs., Ltd.*, No. 98-cv-00629-FB (W.D. Tex.) (contempt motion filed March 28, 2012); *FTC v. Nationwide Connections, Inc.*, No. 06-80180 (S.D. Fla. Sept. 18, 2008) (stipulated final order); *FTC v. Websource Media, LLC*, No. H-06-1980 (S.D. Tex. July 17, 2007) (stipulated final order); *FTC v. Epixtar Corp.*, No. 03-8511 (S.D.N.Y. Nov. 29, 2006) (stipulated final order); *FTC v. Mercury Mktg. of Del., Inc.*, No. 00-3281, 2004 WL 2677177 (E.D. Pa. Nov. 22, 2004); *FTC v. 800 Connect, Inc.*, No. 03-CIV-60150 (S.D. Fla. Feb. 4, 2003) (stipulated final order); *FTC v. Access Resource Servs., Inc.*, No. 02-CIV-60226 (S.D. Fla. Nov. 4, 2002) (stipulated final order); *FTC v.Cyberspace.com, LLC*, No. C00-1806L, 2002 WL 32060289 (W.D. Wash. July 10, 2002), *aff'd,* 453 F.3d 1196 (9th Cir. 2006).

[24] *See, e.g.*, Wise Media Declaration, *supra* note 20, at Exs. FTC 71, FTC 72, FTC 73.

[25] *See, e.g., id.* at Exs. FTC 66, FTC 78; Exhibits PX 12, PX19 to Exhibits in Support of Plaintiff's *Ex Parte* Application for a Temporary Restraining Order (Dkt. #12), *FTC v. Tatto, Inc.*, No. 2:13-cv-08912-DSF-FFM (C.D. Cal. Dec. 5, 2013).

[26] *See, e.g., Inc21.com*, 745 F. Supp. 2d at 1001; Memorandum in Support of Motion for Temporary Restraining Order, at 10-11, *FTC v. Wise Media, LLC*, No. 1:13-cv-1234-WSD (N.D. Ga. Apr. 16, 2013) [hereinafter "Wise Media TRO Memo"]; FTC Roundtable transcript, A. Sege, at 145; *id.*, J. Chilsen, Citizens Utility Board, at 73; *id.*, C. Witteman, California Public Utilities Commission, at 100; *id.*, P. Singer, Office of Texas Attorney General, at 102.

[27] *See* FTC Roundtable transcript, J. Breyault, at 31-32; Brad Tuttle, *Groceries or Mobile Phone? Plenty of Consumers Spend More on the Latter*, TIME MAGAZINE, Sept. 13, 2012, *available at* http://business.time.com/2012/09/13/groceries-or-mobile-phone-plenty-of-consumers-spend-more-on-the-latter/ (noting that nearly half of Americans with mobile phones said their monthly bill comes to $100 or more per month); Anton Troianovski, *Cellphones Are Eating the Family Budget*, WALL ST. J., Sept. 28, 2012, *available at* http://online.wsj.com/article/SB10000872396390444083304578018731890309450.html.

[28] *See, e.g., Cricket Wireless Terms and Conditions of Service*, CRICKET WIRELESS, https://www.cricketwireless.com/terms (last visited July 23, 2014); *Choose a Prepaid Plan. Pick Your Device, View Plan Details*, VERIZON WIRELESS, http://www.verizonwireless.com/b2c/prepay/processPrePayRequest.do?type=ppmonthBASIC (last visited July 23, 2014).

[29] *See, e.g.*, FTC Roundtable transcript, J. Breyault, at 32; *see also id.*, J. Chilsen, at 72-73.

[30] Complaint for Permanent Injunction and Other Equitable Relief, at 7-8, *FTC v. Wise Media, LLC*, No. 1:13-cv-1234-WSD (N.D. Ga. Apr. 16, 2013), *available at* http://www.ftc.gov/sites/default/files/documents/cases/2013/04/130417wisemediacmpt.pdf.

[31] *See* Wise Media TRO Memo, *supra* note 26, at 6-9.

[32] Stipulated Order for Permanent Injunction and Monetary Judgment Against Defendants Brian M. Buckley and Wise Media, LLC, at 4-6, *FTC v. Wise Media, LLC*, No. 1:13-cv-1234-WSD (N.D. Ga. Nov. 22, 2013), *available at* http://www.ftc.gov/sites/default/files/documents/cases/131121wisemediabuckleystip.pdf [hereinafter "Wise Media Settlement"].

[33] Complaint for Permanent Injunction and Other Equitable Relief, at 8-25, *FTC v. Jesta Digital, LLC*, No. 1:13-cv-01272 (D.D.C. Aug. 20, 2013), *available at* http://www.ftc.gov/sites/default/files/documents/cases/2013/08/130821jestacmpt.pdf [hereinafter "Jesta Digital Complaint"].

[34] As discussed in note 128, *infra*, WAP opt-in involves consumers responding to an offer displayed on the mobile web by clicking on a confirmation button from the phone two separate times. This process captures the consumer's phone number without the need for the consumer to enter it manually.

[35] *See* Jesta Digital Complaint, *supra* note 33, at 24.

[36] Complaint for Permanent Injunction and Other Equitable Relief, at 9, *FTC v. Tatto, Inc.*, No. 2:13-cv-08912- DSF-FFM (C.D. Cal. Dec. 5, 2013), *available at* http://www.ftc.gov/sites/default/files/documents/cases/131216bullroarercmpt.pdf.

[37] *See id.*; Memorandum In Support of Plaintiff's Ex Parte Application For Temporary Restraining Order With An Asset Freeze and Other Equitable Relief, And Order to Show Cause Why A Preliminary Injunction Should Not Issue, at 12, *FTC v. Tatto, Inc.*, No. 2:13-cv-08912-DSF-FFM (C.D. Cal. Dec. 5, 2013) [hereinafter "Tatto TRO Memo"].

[38] *See* Stipulated Order for Permanent Injunction and Monetary Judgment Against Defendants Tatto, Inc., Shaboom Media, LLC, Bune, LLC, Mobile Media Products, LLC, Chairman Ventures, LLC, Galactic Media, LLC, Virtus Media, LLC, and Lin Miao, *FTC v. Tatto, Inc.*, No. 2:13-cv-08912-DSF-FFM (C.D. Cal. June 11, 2014), *available at* http://www.ftc.gov/system/files/documents/cases/140613bullroarerstiporder.pdf. The judgment was partially suspended based on defendants' ability to pay, but the defendants that have settled to date have surrendered more than $10 million in assets to be used for consumer redress.

[39] Amended Complaint for Permanent Injunction and Other Equitable Relief, *FTC v. Acquinity Interactive, LLC*, No. 14-60166-CIV (S.D. Fla. June 16, 2014), *available at* http://www.ftc.gov/system/files/documents/cases/140707revenuepathcmpt.pdf.

[40] Complaint for Permanent Injunction and Other Equitable Relief, *FTC v. MDK Media, Inc.*, No. 2:14-cv-05099- JFW-SH (C.D. Cal. July 3, 2014).

[41] *See* T-Mobile Complaint, *supra* note 20, at 11-36.

[42] Public Notice, Fed. Commc'ns Comm'n, Consumer and Governmental Affairs Bureau Seeks to Refresh the Record Regarding "Cramming" (Aug. 27, 2013), *available at* http://transition.fcc.gov/Daily_Releases/Daily_Business/2013/db0827/DA-13-1807A1.pdf [hereinafter "FCC Refresh the Record"].

[43] *See* FTC Reply Comment, *supra* note 4, at 1, 5-7, 12. 44 Press Release, Fed. Commc'ns Comm'n, FCC Investigates Cramming Complaints Against T-Mobile (July 1, 2014), *available at* http://www.fcc.gov/document/fcc-investigates-cramming-complaints-against-t-mobile.

[45] *See* Press Release, S. Comm. on Commerce, Science, and Transp., Rockefeller Vows to Avert Wireless Cramming Scams on Consumers (Mar. 1, 2013), *available at* http://www.commerce.senate.gov/public/index.cfm?p=HearingsandPressReleases&ContentRecord_id=cd0edc13- b355-4d4e-9619-7035329daa1a&ContentType_id=77eb43da-aa94-497d-a73f-5c951ff72372&Group_id=165806cd-d931-4605-aa86-7fafc5fd3536&MonthDisplay =3&YearDisplay=2013; Press Release, S. Comm. on Commerce, Science, and Transp., Rockefeller Questions Billing Aggregators on Wireless Cramming (Mar. 22, 2013), *available at* http://www.commerce.senate.gov/public/index.cfm?p= HearingsandPressReleases&ContentRecord_id=07f5f79c- f6c6-4ca7-a03b-ca76ac5d3962&ContentType_id=77eb43da-aa94-497d-a73f-5c951ff72372&Group_id=165806cd-d931-4605-aa86-7fafc5fd3536&MonthDisplay=3&YearDisplay=2013.

[46] *See, e.g.*, Letter from Sen. John D. Rockefeller IV to Randall Stephenson, March 1, 2013, *available at* http://www.commerce.senate.gov/public/?a=Files.Serve&File_id=2610b9ba-a3d1-43eb-a94f-d7505f354680.

[47] *See* Nat'l Ass'n of Att'y Gen., *Comments for the FTC Mobile Cramming Roundtable* (June 24, 2013), at 1, *available at* http://www.ftc.gov/sites/default/files/documents/public_comments/2013/06/564482-00015-86106.pdf [hereinafter "NAAG Comments"].

[48] *See id.* at 3-4. The Attorneys General also note that some consumers who received text messages suggesting they had been opted in to a service responded "STOP" but were still signed up. *See id.* at 11. Many of these themes mirror landline cramming complaints.

[49] *Id.* at 12.

[50] *See id.* at 7-8. *See also, e.g.*, Assurance of Voluntary Compliance, *In re Verizon Wireless Services LLC & Alltel Commc'ns, LLC*, Case Nos. L08-3-1035 & L08-3-1034 (Fla. Att'y Gen. June 16, 2009), *available at* http://myfloridalegal.com/webfiles.nsf/WF/KGRG-7TAJQ2/$file/VerizonAVC.pdf.

[51] NAAG Comments, *supra* note 47, at 8.

[52] *See* Plaintiff's Original Petition, at 18-24, *State of Texas v. Cellzum.com, LLC*, No. D-1-GV-13-000629 (Travis County, Tex. Dist. Ct. July 11, 2013), *available at* https://www.oag.state.tx.us/newspubs/releases/2013/Cellzum_POP_070813.pdf.

[53] NAAG Comments, *supra* note 47, at 9.

[54] *See* Plaintiff's Original Petition, *State of Texas v. Mobile Messenger U.S. Inc.*, No. D-1-GV-13-001256 (Travis County, Tex. Dist. Ct. Nov. 6, 2013), *available at* https://www.oag.state.tx.us/newspubs/releases/2013/Mobile- Messenger-POP.pdf [hereinafter "Mobile Messenger Complaint"].

[55] Unless otherwise noted, the following discussion refers to cramming by commercial merchants. The FTC has not seen evidence to date of cramming in the context of charitable or political donations.

[56] *See* FTC Roundtable transcript, M. Altschul, CTIA – The Wireless Association, at 24-25; Comments of CTIA – The Wireless Association, FCC CG Docket No. 11-116 (June 25,

2012), at 5-6; Comments of Sprint Nextel Corporation, FCC CG Docket No. 11-116 (Oct. 24, 2011), at 7.

[57] *See* FTC Roundtable transcript, C. Witteman, at 91; Cal. Pub. Util. Comm'n, *Letter to Melanie K. Tiano, S. Comm. on Commerce, Science, and Transp., Re: Inquiry Regarding Wireless California Cramming Complaint Data* (Jan. 31, 2013), *available at* http://apps.fcc.gov/ecfs/document/view;jsessionid=ng1nRV9Wynn82QmDpwMwChl87kFMQhvhcqH5Q21ZLqg b1cJN2Tn1!-1705390101!956499833?id=7022119377 [hereinafter "CPUC Letter"] (percentages extracted from data in the letter). California regulations require wireless carriers to submit monthly data on third-party charges and refunds. *See* CAL. PUB. UTIL., *Final Decision Adopting California Telephone Corporation Billing Rules*, Rulemaking No. 00-02-004 (Nov. 3, 2010), Attach. A at 8 [hereinafter "CPUC Rule"].

[58] *See* Wise Media TRO Memo, *supra* note 26, at 10; Wise Media Declaration, *supra* note 20, at FTC 1 – 000012, 25, Ex. 21.

[59] *See* FTC Roundtable transcript, J. Manis, at 58.

[60] *See FTC v. Commerce Planet, Inc.*, 878 F. Supp. 2d 1048, 1075 (C.D. Cal. 2012); *FTC v. Grant Connect, LLC*, 827 F. Supp. 2d 1199, 1222 (D. Nev. 2011).

[61] *See* NACHA – The Electronic Payments Association, NACHA Operating Rules, Art. II, § 2.17.2 (ODFI Return Rate Reporting) and Appendix 10 §§ 10.2.2, 10.4.3 (Initiation of a Rules Enforcement Proceeding) (2014) [hereinafter "NACHA Operating Rules"]. The 2014 edition of the NACHA Rules is available at www.achrulesonline.org.

[62] *See* FTC Roundtable transcript, J. Bruner, Aegis Mobile, at 82; *id.*, M. Altschul, at 146.

[63] *See* Wise Media Declaration, *supra* note 20, at FTC 1 - 000012-13, 27.

[64] *See id.* at Ex. FTC 21.

[65] *See* FTC Roundtable transcript, M. Altschul, at 146, 152-53; *id.*, M. Niejadlik, at 147-48; *id.*, A. Sege at 149-150; Sprint Nextel Comments, *supra* note 6, at 2, 6; Verizon Wireless Comments, *supra* note 6, at 5.

[66] *See* FTC Roundtable transcript, P. Singer, at 102; Wise Media Declaration, *supra* note 20, at Ex. FTC 69, FTC; JANE KOLODINSKY, CTR. FOR RURAL STUDIES, UNIV. OF VT., MOBILE PHONE THIRD-PARTY CHARGE AUTHORIZATION STUDY (2013), Appendix C at 6-8, *available at* http://www.atg.state.vt.us/assets/files/Mobile%20Phone%20Third- Party%20Charge%20Authorization%20Study.pdf; NAAG Comments, *supra* note 47, at 3-4.

[67] *See* Wise Media TRO Memo, *supra* note 26, at 11-12; FTC Roundtable transcript, J. Breyault, at 55; *id.*, P. Singer, at 102; NAAG Comments, *supra* note 47, at 4.

[68] *See* FTC Reply Comment, *supra* note 4, at 5. These complaints are consistent in their content with hundreds if not thousands of consumer complaints in online message boards. *See*, for example, the many reports of cramming at www.smswatchdog.com, www.textcomplaints.com, and the forums on the carriers' websites. According to the National Association of Attorneys General, several State Attorneys General recently reviewed cramming complaints received by 28 states and found over 750 mobile cramming complaints, predominantly from the last few years. *See* NAAG Comments, *supra* note 47, at 2.

[69] The evidence suggests that consumers who become aware of crammed charges and take the time to complain are far more likely to contact their carrier than government agencies.

[70] Wise Media TRO Memo, *supra* note 26, at 10.

[71] Tatto TRO Memo, *supra* note 37, at 10.

[72] CPUC Letter, *supra* note 57, at 2; *see also* FTC Roundtable transcript, C. Witteman, at 91. The State of Vermont also reports that it has received around two dozen mobile cramming complaints in the last seven years, but it identified hundreds of consumers whose responses indicated that they had been crammed when it proactively surveyed consumers who had incurred third-party charges on their mobile phone bills. FTC Roundtable transcript, K. McCabe, at 12.

[73] *See* KOLODINSKY, *supra* note 66, at 6-7; FTC Roundtable Transcript, K. McCabe, at 12, 28-29.

[74] *See* Citizens Utility Board, *Analysis: Frequency Of Cellphone 'Cramming' Scam Doubles In Illinois, CUB Concerned Wireless Customers Targeted As Landline Laws Tighten* (Dec. 4, 2012), *available at* http://www.citizensutilityboard.org/newsReleases20121204_CellphoneCramming.html.

[75] CTIA – The Wireless Ass'n, *Re: "Mobile Cramming" Roundtable, Project No. P134830*, (June 24, 2013), at 1, *available at* http://www.ftc.gov/policy/public-comments/comment-564482-00016. Even if the percentage of charges that were unauthorized was somewhat lower than those reported in the Vermont study, this would still point to a substantial volume of unauthorized third-party billing.

[76] Comments of CTIA – The Wireless Ass'n, FCC CG Docket No. 11-116 (June 25, 2012), at 3-6.

[77] *See* Verizon Wireless Comments, *supra* note 6; Sprint Nextel Comments, *supra* note 6. Specific industry approaches are discussed in detail in Part IV, *infra*.

[78] *See* FTC Roundtable transcript, J. Manis, at 15.

[79] *See, e.g.*, FTC Reply Comment, *supra* note 4, at 6-7; Consumers Union, *Comments for the FTC Mobile Cramming Roundtable* (June 24, 2013), at 2-3, *available at* http://www.ftc.gov/sites/default/files/documents/public_comments/2013/06/564482-00014-86099.pdf.

[80] *See* Exhibits in Support of Motion for Temporary Restraining Order, *FTC v. Inc21.com Corp.*, No. 3:10-cv-00022-WHA (N.D. Cal. Jan. 5, 2010), at Ex. 1, at 4.

[81] *See FTC v. Inc21.com Corp.*, 745 F. Supp. 2d at 1001.

[82] *See* Press Release, Fed. Trade Comm'n, FTC Returns More Than $5.4 Million to Victims of Massive Cramming Scam (Sept. 30, 2013), *available at* http://www.ftc.gov/opa/2013/09/inc21.shtm.

[83] *See* Wise Media Declaration, *supra* note 20, at FTC 1 - 000013-14, 30; Wise Media Settlement, *supra* note 32, at 6. The judgment was partially suspended based on defendants' ability to pay.

[84] FED. TRADE COMM'N STAFF, CONSUMER FRAUD IN THE UNITED STATES: AN FTC SURVEY (2004), *available at* http://www.ftc.gov/sites/default/files/documents/reports/consumer-fraud-united-states-ftc- survey/040805confraudrpt.pdf.

[85] *See Inc21.com Corp.*, 745 F. Supp. 2d at 1001.

[86] *See* FTC Roundtable transcript, A. Sege, at 145; *id.*, C. Witteman, at 100; *id.*, P. Singer at 86-87; *see also* NAAG Comments, *supra* note 47, at 3, 6.

[87] *See* Wise Media TRO Memo, *supra* note 26, at 6.

[88] *See* Graphic, "Excerpts from an actual T-Mobile bill," *available at* http://www.ftc.gov/system/files/attachments/press-releases/ftc-alleges-t-mobile-crammed-bogus-charges- customers-phone-bills/tmobile-samplebill.pdf.

[89] *Supra* Part III.A, at 10.

[90] *Id.*

[91] *See* FTC Roundtable transcript, L. Bryenton, Competition Bureau, Canada, at 49. *See also* Press Release, Competition Bureau (Canada), Competition Bureau Sues Bell, Rogers and Telus for Misleading Consumers: Bureau Seeks Customer Refunds and $31 Million in Penalties (Sept. 14, 2012), *available at* http://www.competitionbureau.gc.ca/eic/site/cb-bc.nsf/eng/03498.html.

[92] *See* PhonepayPlus, *Comments for the FTC Mobile Cramming Roundtable* (July 16, 2013), at 2, *available at* http://www.ftc.gov/os/comments/mobilecramming/564482-00017-86205.pdf.

[93] *See* Wayne Flower, *False Text Message Racket Cashes in on Mobile Phone Charges*, HERALD SUN, June 23, 2013, *available at* http://www.heraldsun.com.au/news/law-order/false-text-message-racket-cashes-in-on-mobile- phone-charges/story-fni0fee2-1226668456799; Katia Moskvitch, *Hidden Mobile Charges that Could Be Buried in Your Bill*, BBC NEWS, Sept. 6, 2012, *available at* http://www.bbc.co.uk/news/technology-19402398.

[94] LONDON ACTION PLAN & M^{3}AAWG, BEST PRACTICES TO ADDRESS ONLINE AND MOBILE THREATS (Oct. 15, 2012), *available at* http://www.maawg.org/sites/maawg/files/news/M3AAWG_LAP_Best_Practices_to_Address_Online_and_Mobil e_Threats_0.pdf.

[95] *Id.* at 36.

[96] AUSTRALIAN COMPETITION & CONSUMER COMM'N, MOBILE PREMIUM SERVICES: MEETING THE CHALLENGES (Sept. 18, 2012), at 1, 5-8, *available at* http://www.accc.gov.au/system/files/Mobile%20Premium%20Services%20-%20Case%20Study.pdf.

[97] *Id.* at 8.

[98] *Id.* at Figure 2.

[99] *See* FTC Roundtable transcript, M. Altschul, at 34-35, 171-72, 177; Verizon Wireless Comments, *supra* note 6, at 4; Sprint Nextel Comments, *supra* note 6, at 4.

[100] Some carriers also provide the ability to block charges from a specific third party's services. *See* FTC Roundtable transcript, M. Altschul, at 172, 177.

[101] *See id.*, at 34-35; Verizon Wireless Comments, *supra* note 6, at 5; Sprint Nextel Comments, *supra* note 6, at 4; FTC Roundtable transcript, C. Witteman, at 92; California Public Utilities Commission Staff, Effectiveness of the Cramming Rules in Decision 10-10-034 in Protecting California Consumers from Unauthorized Charges on Their Phone Bills, and Related Developments in the Wireless Industry 16-17 (2014), *available at* http://www.cpuc.ca.gov/PUC/hottopics/2Telco/01_June_2014_Cramming_Report.htm; *see also* Nat'l Consumers League, et al., *Comments for the FTC Mobile Cramming Roundtable* (June 7, 2013) at 8, *available at* http://www.ftc.gov/sites/default/files/documents/public_ comments/2013/06/564482-00013-86076.pdf [hereinafter "Consumer Groups FTC Comment"]. Additionally, the State of California requires phone companies to disclose to consumers that they can block all third-party charges on their mobile bills. *See* FTC Roundtable transcript, C. Witteman, at 90, 92; CPUC Rule, *supra* note 57, Attach. A at 4. From the record, it is not clear whether all carriers currently have this option and make these disclosures.

[102] *See* Comments of Center for Media Justice, Consumer Action, Consumer Federation of America, Consumers Union, National Consumer Law Center – on Behalf of Its Low-Income Clients, and National Consumer League, FCC CG Docket No. 11-116, at 18-19 (June 25, 2012), *available at* http://apps.fcc.gov/ecfs/document/view?id=7021977710 [hereinafter "Consumer Groups FCC Comment"]. The FCC is considering whether to modify its Truth-in-Billing rules to include a default block on all third-party charges. *See* FCC Refresh the Record, *supra* note 42. This would mean that consumers would have to affirmatively opt-in to third-party charges before such charges could be placed on their mobile phone bills.

[103] FTC Roundtable transcript, K. McCabe, at 61; A. Sege, at 130.

[104] *See id.*, M. Altschul, at 171, 177.

[105] *See* FTC Reply Comment, *supra* note 4, at 12.

[106] In earlier comments to the FCC, the Commission supported a ban or default block on third-party billing for *landline* billing, but stated that additional information was needed to determine whether a default block was appropriate in the mobile context. *See* FTC Reply Comment, *supra* note 4, at 3, 11-12. The Commission noted that, in contrast to landline third-party billing, the mobile billing platform has been used for some legitimate purposes – including charitable giving – and it is a potential platform for consumers to fund mobile payments for desired services. Indeed, some consumer advocacy groups have agreed that there is legitimate commerce occurring through mobile billing. *See* Consumer Groups FTC Comment, *supra* note 101, at 9. *See also* FTC Roundtable transcript, D. Derakhshani, Consumers Union, at 174. Based on the current evidence regarding authorized third-party carrier billing, as well as the potential costs and benefits of blocking third-party charges by default, staff does not believe that a default block on mobile third-party billing is warranted at this time.

[107] *See* FTC Roundtable transcript, M. Altschul, at 171-72, 177.

[108] *See* Report and Order and Further Notice of Proposed Rulemaking, Docket No. 11-116, at 59 (Apr. 27, 2012), *available at* http://hraunfoss.fcc.gov/edocs_public/attachmatch/FCC-12-42A1.pdf [hereinafter "FCC Report and Order"]; 47 C.F.R. § 64.2401(f).

[109] *See* FTC Roundtable transcript, M. Altschul, at 171-72, 177.

[110] Carriers should ensure that their blocking services provide the protections they are purported to provide. The Senate Commerce Committee's report on landline cramming found that consumers often incurred additional crammed charges on their landline telephone bills even after they requested the blocking of third-party charges on their landline telephone accounts. *See* S. COMMERCE COMM. CRAMMING REPORT, *supra* note 22, at 33-35, App. A. Mobile carriers should take steps to ensure that the blocking options they offer consumers are effective.

[111] The MMA periodically revised and published these as the "U.S. Consumer Best Practices for Messaging." The most recent version (7.0) was published on October 16, 2012. *See* MOBILE MARKETING ASS'N, U.S. CONSUMER BEST PRACTICES FOR MESSAGING VERSION 7.0 (Oct. 16, 2012), *available at* http://www.mmaglobal.com/uploads/Consumer-Best-Practices.pdf. However, the MMA has stated that it will make no further revisions. *See* FTC Roundtable transcript, C. Frey, Mobile Marketing Association, at 76, 106-07.

[112] CTIA – THE WIRELESS ASSOCIATION, COMPLIANCE ASSURANCE SOLUTION: MOBILE COMMERCE COMPLIANCE HANDBOOK (November 4, 2013), *available at* http://www.wmcglobal.com/assets/ctia-mobile- commerce-compliance-handbook-v-1-3.pdf [hereinafter "CTIA Handbook"].

[113] *See, e.g.*, FTC Roundtable transcript, P. Singer, at 108-09.

[114] *See id.*, at 85-86; FTC Roundtable, *PowerPoint Presentation re: State of Texas v. Eye Level Holdings*, at 6, PowerPoint slides *available at* http://www.ftc.gov/sites/default/files/documents/public_events/mobile-cramming- roundtable/state-texas-v-eye-level-holdings-et-al.ppt [hereinafter "Eye Level Holdings Presentation"].

[115] *See* FTC Roundtable transcript, P. Singer, at 86; Eye Level Holdings Presentation, *supra* note 114, at 8.

[116] *See* FTC Roundtable transcript, C. Frey, at 76.

[117] *See, e.g.,* Wise Media TRO Memo, *supra* note 26, at 12-13.

[118] *See, e.g.*, Verizon Wireless Comments, *supra* note 6, at 2; Sprint Nextel Comments, *supra* note 6, at 5; FTC Roundtable transcript, M. Altschul, at 25-26, 44-45; *id.*, J. Bruner, at 81.

[119] *See id.*, J. Bruner, at 81, 96-97, 111; Verizon Wireless Comments, *supra* note 6, at 3-4; Sprint Nextel Comments, *supra* note 6, at 5.

[120] *See* FTC Roundtable transcript, A. Sege, at 141.

[121] *See id.*, P. Singer, at 87; *id.*, J. Bruner, at 97; Order and Preliminary Injunction, *Cellco P'ship v. Hope*, No. CV11-0432-PHX-DGC (D. Ariz. May 11, 2011), at 4-6.

[122] *See* FTC Roundtable transcript, J. Bruner, at 97-98.

[123] *See id.*, J. Bruner, at 104; *id.*, J. Breyault, at 13; Consumer Groups FTC Comment, *supra* note 101, at 9-10.

[124] *See* FED. TRADE COMM'N STAFF, REVISED .COM DISCLOSURES: HOW TO MAKE EFFECTIVE DISCLOSURES IN DIGITAL ADVERTISING 10 (2013), *available at* http://www.ftc.gov/sites/default/files/attachments/press- releases/ftc-staff-revises-online-advertising-disclosure-guidelines/130312dotcomdisclosures.pdf.

[125] Although the MMA and CTIA guidelines already purport to require pricing and payment information, the examples cited in the guidelines do not always make clear that it is the consumer's phone bill that will be charged. *See* MOBILE MARKETING ASS'N, *supra* note 111, at 23 (Best Practice 2.3-5) ("All advertising must clearly disclose the subscription term, billing interval and information on how the charges will be applied (*i.e.*, that the charges will be billed on the customer's wireless phone bill or deducted from the customer's prepaid balance)."); *but compare* CTIA Handbook, *supra* note 112, at 19-20, *and* MOBILE MARKETING ASS'N, *supra* note 111, at 38, 40. Ensuring that the consumer understands the

billing mechanism is consistent with rules adopted by the Commission in other contexts. *See* FTC Telemarketing Sales Rule, 16 C.F.R. § 310.4(a)(7)(i)-(ii).

[126] *See, e.g.*, Fed. Deposit Ins. Corp., FDIC Supervisory Approach to Payment Processing Relationships With Merchant Customers That Engage in Higher-Risk Activities, FIL-43-2013 (Sept. 27, 2013), *available at* http://www.fdic.gov/news/news/financial/2013/fil13043.pdf; Fed. Deposit Ins. Corp., Payment Processor Relationships Revised Guidance, FIL-3-2012 (January 31, 2012), *available at* http://www.fdic.gov/news/news/financial/2012/fil12003.pdf; Office of the Comptroller of the Currency, OCC Bulletin 2013-29 (Oct. 30, 2013), *available at* http://occ.gov/news-issuances/bulletins/2013/bulletin-2013-29.html; Office of the Comptroller of the Currency, OCC Bulletin 2006-39 (Sept. 1, 2006), *available at* http://www.occ.gov/news-issuances/bulletins/2006/bulletin-2006-39.html. Banks also require their payment processor customers to comply with KYC procedures for merchant-customers.

[127] *See, e.g.*, VISA, ACQUIRER BEST PRACTICES CONTINUITY MERCHANTS (Aug. 26, 2010), *available at* https://usa.visa.com/download/merchants/bulletin_ acquirer_continuity_ best_practices.pdf; NACHA Operating Rules, *supra* note 61, at § 2.2.3; *see also* VISA, VISA E-COMMERCE MERCHANTS' GUIDE TO RISK MANAGEMENT 24-25 (2008), *available at* https://usa.visa.com/download/merchants/visa-risk-management-guide-ecommerce.pdf.

[128] *See* FTC Roundtable transcript, A. Sege, at 128; *id.*, J. Bruner, at 113. Common forms of double opt-in include:

1. Mobile phone opt-in – a consumer sends a text message with a keyword to a short code, receives a confirmation text, and then sends a second text message to confirm the transaction;
2. Web opt-in – a consumer first enters his or her number on a website, then receives a text with a PIN, which the consumers enters back into the website;
3. WAP (wireless access protocol) – a consumer responds to an offer on his or her phone (displayed through the mobile web), and clicks on a confirmation button on the phone two separate times; and
4. IVR (interactive voice response) – a consumer opts-in over the phone through a toll-free number and confirms the transaction twice. *See* FTC Roundtable transcript, J. Sizer, Aegis Mobile, at 122-23.

[129] *See* Wise Media TRO Memo, *supra* note 26, at 12-13; FTC Roundtable transcript, J. Bruner, at 82; *id.*, M. Altschul, at 146.

[130] *See* Wise Media Declaration, *supra* note 20, at FTC 1 - 000012-13, 27.

[131] *See* FTC Roundtable transcript, M. Tiano, S. Comm. on Commerce, Science, and Transp., at 164; NAAG Comments, *supra* note 47, at 11-12; Consumer Groups FCC Comment, *supra* note 102, at 10-15.

[132] *See supra* Part III.A, at 11, 14.

[133] *See* FTC Roundtable transcript, M. Altschul, at 39, 62; *id.*, J. Manis, at 27.

[134] *See id.*, M. Altschul, at 39, 62; Verizon Wireless Comments, *supra* note 6, at 3. At least one major aggregator indicated that it will require all merchants to use the aggregator to host the opt-in records. *See* FTC Roundtable transcript, A. Sege, at 167-68. The State of Texas, however, recently sued that aggregator for allegedly "orchestrating and facilitating" a cramming scheme. *See* Mobile Messenger Complaint, *supra* note 54, at 5.

[135] *See* FTC Roundtable transcript, J. Manis, at 27.

[136] *See id.*, M. Niejadlik, at 136; *id.*, J. Greenwell, at 65.

[137] *See* FTC Roundtable transcript, J. Manis, at 26-27.

[138] *See id.*, J. Greenwell, at 65.

[139] *See* FTC Roundtable transcript, C. Witteman, at 114-15.

[140] *See* NAAG Comments, *supra* note 47, at 12.

[141] CPUC Rule, *supra* note 57, Attach. A at 4; *see also* FTC Roundtable transcript, C. Witteman, at 90.

[142] Nat'l Ass'n of State Util. Consumer Advocates, *Urging State Legislatures to Prohibit the "Cramming" of Unauthorized Charges Onto Consumer Telephone Bills* (June 11, 2013), at 2, *available at* http://nasuca.org/2013-01-urging-state-legislatures-to-prohibit-the-cramming-of-unauthorized-charges-onto-consumer-telephone-bills- and-proposing-a-statute-to-solve-the-problem/.
[143] *Id.* at 7.
[144] *See* Wise Media Declaration, *supra* note 20, at Ex. FTC 21.
[145] *See supra* Part III.B.1, at 16.
[146] *See supra* Part III, at 10, 20.
[147] *See* FTC Roundtable transcript, P. Singer, at 86-87.
[148] *See, e.g.*, Wise Media Declaration, supra note 20, at Exs. FTC 66, FTC 71, FTC 72, FTC 73; Wise Media TRO Memo, *supra* note 26, at 6 n.2.
[149] T-Mobile Complaint, *supra* note 20, at 13.
[150] *See* FTC Roundtable transcript, P. Singer, at 118; i*d.*, M. Niejadlik, at 155-56; *id.*, K. McCabe, at 12, 28.
[151] *See* FTC Roundtable transcript, L. Follansbee, Fed. Commc'ns Comm'n, at 133; FCC Report and Order, *supra* note 108, at 6; 47 C.F.R. § 64.2401(b).
[152] *See* 47 C.F.R. § 64.2401(a); FCC Report and Order, *supra* note 108, at 20.
[153] *See* FTC Roundtable transcript, L. Follansbee, at 133-34; FCC Report and Order, *supra* note 108, at 20-21, 53; FCC Refresh the Record, *supra* note 42, at 1.
[154] FTC Reply Comment, *supra* note 4, at 12.
[155] *See* Sprint Nextel Comments, *supra* note 6, at 6; Verizon Wireless Comments, *supra* note 6, at 5.
[156] *See* FTC Roundtable transcript, M. Altschul, at 60.
[157] *Id.*, M. Altschul, at 185-86.
[158] FCC Workshop, *Bill Shock and Cramming* (Apr. 17, 2013), G. Reynolds, United States Telecom Association, at 99:15, *recording available at* http://www.fcc.gov/events/workshop- bill-shock-and-cramming.
[159] *See* FTC Roundtable transcript., M. Tiano, at 186-87 (discussing wireline cramming); *id.*, J. Breyault, at 55; *id.*, P. Singer 101-02, 116-17; Consumer Groups FTC Comment, *supra* note 101, at 7.
[160] *See* FTC Roundtable transcript, P. Singer, at 116-117; Wise Media TRO Memo, *supra* note 26, at 11-12.
[161] *See* FTC Roundtable, K. McCabe, at 61; KOLODINSKY, *supra* note 66, App. C at 6-8.
[162] *See* NAAG Comments, *supra* note 47, at 3-4.
[163] *See id.*
[164] FTC Roundtable transcript, J. Chilsen, at 121; Consumer Groups FCC Comment, *supra* note 102, at 21.
[165] FTC Roundtable transcript, J. Chilsen, at 121.
[166] *Id.*, M. Altschul, at 152.
[167] FTC Reply Comment, *supra* note 4, at 12.
[168] FTC MOBILE PAYMENTS REPORT, *supra* note 6, at 5-8.
[169] *See* 12 C.F.R. §§ 1026.12, 1026.13.
[170] *See* 12 C.F.R. §§ 1005.6, 1005.11.
[171] The Senate Commerce Committee's report on landline cramming discusses a customer survey that a landline carrier conducted via email when it suspected that a third party had engaged in landline cramming. None of the consumers who responded to the survey reported that they had authorized the third-party charge at issue. None of the first twelve respondents to the survey had complained to the landline carrier about the unauthorized third- party charge before the landline carrier conducted the survey, and it is not known whether any of the other survey respondents had complained to the carrier before receiving the survey either. S. COMMERCE COMM. CRAMMING REPORT, *supra* note 22, at 37.

[172] *See T-Mobile Will No Longer Allow Third Parties to Bill Customers for Premium SMS Services*, T-MOBILE, http://newsroom.t-mobile.com/news/t-mobile-announces-new-program-related-to-premium-sms-charges.htm (last visited July 23, 2014) ("Despite protections and processes put in place by T-Mobile and the industry, not all Premium SMS vendors have acted responsibly."); Pamela Prah, *'Cramming' phone scams targeted by state attorneys general*, USA TODAY, Dec. 19, 2013, *available at* http://www.usatoday.com/story/news/nation/2013/12/19/stateline-cellphone-cramming/4129561/ (Verizon Wireless statement that it was in the process of winding down its premium messaging business, in part, because of "recent allegations that third parties have engaged in improper conduct in providing premium messaging services to our customers.").

In: Cramming and Third Party Billing …
Editor: Christopher Rye
ISBN: 978-1-63463-121-1

Chapter 3

STATEMENT OF JOHN R. THUNE, RANKING MEMBER, SENATE COMMITTEE ON COMMERCE, SCIENCE, AND TRANSPORTATION. HEARING ON "CRAMMING ON WIRELESS PHONE BILLS: A REVIEW OF CONSUMER PROTECTION PRACTICES AND GAPS"*

I would like to thank Chairman Rockefeller for holding this hearing to discuss unauthorized charges on mobile phone bills and the findings of his wireless cramming investigation. I commend the chairman and his staff for shining a light on these abuses. I understand that Senator Blumenthal has graciously offered to serve in his place as chair.

Mobile payments are a growing way for consumers to pay for goods and services. Third-party billing is one way that consumers can take advantage of new technologies and customer conveniences. There are legitimate uses for this manner of billing. For instance, consumers can provide money to charity, support a political cause, or download the newest song or app, and bill the purchase directly to their mobile telephone bill.

* This is an edited, reformatted and augmented version of a statement presented July 30, 2014 before the Senate Committee on Commerce, Science, and Transportation.

Yet, despite industry efforts to implement protections and state and federal regulations in place to prevent cramming, unscrupulous actors have been able to game the system to take advantage of third-party phone billing.

Of course, cramming is not a new phenomenon.

In the late 1990s, Congress devoted a lot of time and attention to the issue of cramming on *landline* phone bills. This committee held a hearing on that issue again in 2011, highlighting the chairman's investigation into cramming on landline phone bills and demonstrating the persistence of the problem.

Some states have enacted laws to limit third-party billing in an effort to prevent cramming on landline phones. And most of the major phone carriers have ended most types of third-party billing on landline phone bills. More recently, however, concerns have been raised about fraud on *wireless* phone bills – the topic of today's hearing and the chairman's more recent investigation.

There are three key parties involved in placing third party charges on consumers' wireless phone bills: the third-party content provider, the billing aggregator, and the phone carrier. From what I have seen, there are some content providers and even some aggregators that appear to be bad actors, but *all* of the parties involved could do more to protect consumers from cramming.

While cramming has been identified as a problem, it has been challenging to accurately measure how many consumers have been affected by cramming. I appreciate that the wireless carriers and their association, CTIA, have taken a number of actions to prevent cramming of third-party charges on wireless phone bills. Significantly, this past November, the carriers decided to end most so-called "premium" SMS programs, which billed customers for text messages related to topics like daily horoscopes and sports alerts. In addition, at least one carrier has recently decided to end browser-based direct carrier billing. These steps show the carriers treat this issue seriously, but we'll be asking whether they should do more.

At the same time, it's important to underscore the extraordinary innovation and economic dynamism in the wireless communications space. The owners of the approximately 188 million smartphones in this country spend more time with their mobile devices each day than they do going online with a laptop or PC. While we must strive to protect consumers from fraud, we must also make sure that we do so in a way that does not stifle innovation. I look forward to hearing from CTIA, who is here today representing the wireless carriers, to discuss how the industry is working to address these issues.

I also look forward to hearing from FTC Commissioner McSweeny, who is here for the first time since her confirmation, Mr. LeBlanc of the FCC, and Attorney General Sorrell. The FTC, FCC, and state attorneys general play a key role in fighting cramming with their law enforcement efforts and by educating consumers about carrier billing. I also would like to thank the South Dakota Public Utilities Commissioners and the South Dakota Attorney General, Marty Jackley, for their work in this area to better protect consumers.

One recent government survey found that the Midwest is the most wireless connected region of the country, with 44 percent of Midwesterners living in cellphone-only homes. This underscores the importance to my constituents of addressing wireless cramming.

This hearing presents a good opportunity to recognize the good that everyone at the witness table is already doing to combat cramming. Industry, Congress, federal agencies, and state attorneys general all need to continue to work together on this issue to ensure that consumers are informed and protected against bad actors.

Thank you to our witnesses for appearing today. I look forward to your testimony.

In: Cramming and Third Party Billing … ISBN: 978-1-63463-121-1
Editor: Christopher Rye

Chapter 4

Statement of Terrell McSweeny, Commissioner, Federal Trade Commission. Hearing on "Cramming on Wireless Phone Bills: A Review of Consumer Protection Practices and Gaps"*

I. Introduction

Chairman Rockefeller, Ranking Member Thune, and members of the Committee, my name is Terrell McSweeny, and I am a Commissioner at the Federal Trade Commission ("FTC" or "Commission").[1] I appreciate this opportunity to appear before you today to discuss the Commission's experience addressing mobile cramming. I am pleased to be testifying alongside my partner at the Federal Communications Commission, with which the FTC has worked collaboratively to combat the problem of mobile cramming. I also would like to commend this Committee and you, Mr. Chairman, for the work you have done to investigate and address this important consumer protection issue.

Mobile cramming is the act of placing unauthorized third-party charges on mobile phone accounts. It occurs when consumers are signed up and billed for third-party services, such as ringtones and recurring text messages containing

* This is an edited, reformatted and augmented version of a statement presented July 30, 2014 before the Senate Committee on Commerce, Science, and Transportation.

trivia or horoscopes, without the consumers' knowledge or consent. Companies that place crammed charges sometimes obtain consumers' phone numbers without any contact with consumers. Other times, these entities use deceptive means to obtain consumers' mobile phone numbers – such as in connection with offering free prizes – and then begin charging their phone accounts for recurring third-party charges for purported services unrelated to the offer. These unauthorized charges often appear buried in phone bills and have generic descriptors such as "usage charges." As a result, many consumers do not notice the charges or do not understand that they are unrelated to their phone service. Moreover, some consumers have prepaid accounts and do not receive bills at all, while others auto-pay their bills and therefore may not routinely inspect them. And many consumers do not even receive the services for which they are being charged.

Mobile cramming is a significant problem that threatens to undermine confidence in the developing payment method known generally as "carrier billing," which offers consumers the opportunity to charge goods and services to their mobile phone accounts. As stakeholders have noted, carrier billing of third-party charges may be particularly beneficial for unbanked and underbanked consumers. Additionally, consumers have used text messages to donate funds to a charitable organization, with the charge placed on their mobile phone account. As carrier billing has developed, however, fraud has become a significant problem for consumers.

For the past two decades, one of the Commission's top priorities has been ensuring that consumer protections keep pace with technological developments, including emerging mobile products and services, while encouraging innovations that benefit consumers and businesses. In the past few years the Commission has focused on mobile cramming as a key consumer protection issue.[2] Among other things, since the spring of 2013, the Commission has brought five mobile cramming cases against merchants, resulting in substantial monetary judgments.[3] And, earlier this month, the Commission filed its first action against a telecommunications company, T-Mobile USA, for mobile cramming.[4] These actions all reinforce the basic principle that a company must obtain a consumer's express, informed consent before placing charges on their bills – which applies to the mobile environment just as it does to brick-and-mortar companies.

In addition to its enforcement work, the Commission has recommended the adoption of certain baseline consumer protections,[5] encouraged public dialogue among industry stakeholders through a public roundtable in May 2013,[6] and, just this week, authorized the release of a Bureau of Consumer

Protection staff report providing additional information about mobile cramming and discussing recommended approaches to address it.[7]

This testimony begins with an overview of the Commission's and this Committee's work to address landline cramming, which has provided the foundation for the Commission's recent efforts to address mobile cramming. The testimony then discusses publicly available evidence regarding the scope of the mobile cramming problem, and the Commission's recent enforcement actions to combat it. Finally, the testimony discusses the recommendations in the FTC staff report released this week.

II. Landline Cramming

As this Committee has recognized, the issue of unauthorized third-party billing on landline phone bills has been a problem for well over a decade. The Committee's investigation and 2011 staff report have played critical roles in illuminating this important consumer protection issue. Indeed, the Committee's staff report estimated that landline cramming has likely cost consumers billions of dollars.[8]

The FTC has brought more than 30 enforcement actions under Section 5 of the FTC Act to halt landline cramming practices and provide restitution to consumers.[9] These cases have resulted in tens of millions of dollars in refunded charges and stringent court orders to prevent future cramming violations. Over the years, the FTC also has worked closely with federal and state partners, including State Attorneys General and the Federal Communications Commission, to combat the problem, and has engaged in consumer and business education to raise awareness about the issue. In addition, the FTC has sought input on the problem from industry participants, consumer groups, and other stakeholders, including by holding a workshop devoted to cramming in 2011.[10] Based on this multi-faceted experience, the FTC has advocated a number of measures to address landline cramming.[11]

III. FTC Enforcement Actions

Over the past few years, it has become apparent that unauthorized third-party charges were appearing not only on landline bills but also on mobile accounts. The use of mobile devices has grown so rapidly that, according to industry, mobile devices now outnumber people in the United States.[12] Building on its experience in the landline arena, the Commission has looked closely at how cramming has spread to mobile accounts. The Commission devoted a portion of the FTC's 2011 cramming workshop to the topic of mobile cramming, filed a comment in an FCC proceeding in July 2012 recommending certain baseline consumer protections,[13] and held a separate roundtable in May 2013 specifically to address mobile cramming.[14] FTC staff also addressed the issue in its April 2013 report on mobile payments.[15] Further, this week, the Commission released a staff report on mobile cramming recommending best practices for industry to prevent and remedy mobile cramming.[16]

As noted above, since the spring of 2013, the Commission also has brought six enforcement actions to prevent mobile cramming and provide restitution for injured consumers. Thus far, the Commission has obtained strong relief in the three actions that have been fully or partially resolved:

- *Tatto, Inc. & Bullroarer, Inc.* In this case, the FTC alleged that a widespread mobile cramming operation engaged in deceptive and unfair practices, for example by running web advertisements that promised consumers a chance to win prizes such as free Justin Bieber tickets and then solicited their phone numbers.[17] Consumers did not receive the Justin Bieber tickets, but rather, as the Commission has alleged, it is likely that consumers were signed up for the defendants' subscription plans.[18] The primary corporate defendants and their operator have agreed to a partially suspended judgment of more than $150 million.[19]
- *Jesta Digital, LLC.* In this case, the FTC alleged that the defendant lured consumers into purchasing a monthly subscription for ringtones using deceptive virus scan ads.[20] According to the complaint allegations, some consumers saw banner ads on their mobile devices while playing a popular mobile app that falsely claimed a virus had been detected. Clicking on the ad led to a screen with a button stating "Get Now" above the phrase "Protect your Android [phone] today." Consumers who clicked "Get Now," and then a button on a subsequent page marked "Subscribe," were then subscribed to the $9.99 per month ringtone subscription plan, though the nature and cost of the subscription were never adequately disclosed. Indeed, some consumers were subscribed even if they clicked on parts of the screen other than the "subscribe" button. Moreover, if consumers

actually attempted to subscribe and download Jesta's so-called anti-virus software to their mobile devices, the download often failed. To obtain consumers' purported authorization for the charges, Jesta used a process known as WAP or Wireless Access Protocol billing,[21] which captures consumers' phone numbers from a mobile device. Thus, consumers never even entered their phone numbers prior to being billed.[22] Under the terms of the settlement, the company must provide refunds to injured consumers and pay an additional $1.2 million to the FTC.[23]

- *Wise Media LLC.* The FTC filed suit in April 2013 against the merchant Wise Media, LLC, which purported to sell recurring subscriptions to text message services providing "love tips," horoscopes, diet tips, and similar kinds of "alerts" for $9.99 a month.[24] The company claimed that consumers signed up for the services by entering their information into websites, receiving PIN codes by text messages, and inputting the PINs into the websites. The FTC alleged that many consumers did not notice the charges, which were often buried in their phone bills, including, in at least one consumer's case, on page 18 of the consumer's bill.[25] Consumers who discovered the charges widely reported that they had never heard of Wise Media or signed up for the services; the FTC alleged that consumers were simply billed without authorization.[26] In November 2013, a court entered a stipulated order with a judgment for more than $10 million and a ban that prohibits Wise Media from placing charges on mobile phone bills altogether.[27]

The Commission is litigating two similar actions against content providers. In *FTC v. Acquinity Interactive, LLC*, the Commission alleges that crammers sent text messages promising free $1,000 gift cards and iPads as a way to deceive consumers into "confirming" their phone number and entering PINs on a website; this resulted in consumers being signed up for unwanted premium text messaging services and incurring charges of $9.99 per month on their mobile phone accounts.[28] In another case, against MDK Media, Inc., the Commission alleges that a content provider similarly used the lure of "free" gift cards to collect consumers' phone numbers and crammed consumers for subscription services such as horoscope alerts.[29]

Earlier this month, the Commission filed suit against T-Mobile USA, alleging that it unlawfully charged consumers for unauthorized monthly text message subscriptions offered by third-party merchants.[30] The complaint alleges that T-Mobile deceptively described these charges on its phone bills in a manner that made it difficult for consumers to discover them. For example, T-Mobile's online bill summaries lumped third-party charges into a line item labeled "Use Charges" that could include charges for both T-Mobile's own text services and for third-party charges.[31]

Additionally, according to the complaint, T-Mobile continued to charge consumers even after becoming aware of telltale signs that the charges were unauthorized. The complaint alleges that T-Mobile's own internal documents showed that consumers increasingly were calling T-Mobile to complain about unauthorized third-party charges. It also alleges that large numbers of consumers sought refunds and the refund rate – the ratio of refunds to charges billed for a particular period of time such as a month – for some subscriptions was as high as 40% in some months. Further, the complaint states that T-Mobile continued to bill consumers for charges from third-party merchants for years after those merchants were the subject of law enforcement or other legal action for cramming, and after news articles and industry alerts detailed cramming behavior and other deceptive behavior by those merchants. On the same day the FTC filed its complaint, the FCC announced that it had opened its own investigation into T-Mobile's practices in regard to cramming.[32]

A number of lessons can be drawn from these actions, as well as the enforcement actions brought by our state law enforcement partners.[33] First, many entities have been able to cram charges onto mobile phone accounts using similar practices, and the amount of money at stake has been substantial. The Wise Media, Jesta Digital, and Tatto/Bullroarer cases alone involved settlements totaling more than $160 million.

Second, the level of consumer complaints and refund requests has understated the overall harm. Carriers have received a large number of complaints and refund requests related to third-party charges on mobile accounts, but the evidence indicates that many consumers do not notice the unauthorized charges, which often are buried in their mobile phone bills and, as alleged in the T-Mobile matter, appear under non-descriptive headers mixed in with charges for phone services.[34] Further, consumers with prepaid mobile phone accounts do not receive a bill at all; unauthorized charges are simply deducted from their available balance of minutes.

Third, even when consumers notice unauthorized charges and have requested refunds, they have reported difficulties obtaining refunds from carriers. Many complain that carriers refuse to give more than two months' worth or other limited amounts of refunds, even if consumers learn that crammed charges have appeared on their bills for longer periods of time.[35] In other instances, carriers have told consumers to contact the merchant for a refund, a request that the merchant often denies.[36]

IV. Staff Recommendations on Best Practices to Address Mobile Cramming

The Commission has advocated certain baseline consumer protections to combat mobile cramming, and the staff report released this week provides staff's additional recommendations for industry best practices. Stakeholders in the mobile billing industry generally have relied on a set of voluntary guidelines to attempt to address cramming, but as demonstrated above, these have not been effective in stopping cramming.[37]

In making its recommendations, Commission staff considered how the mobile carrier billing industry has evolved. Until recently, the dominant type of carrier billing has been "Premium SMS" billing. Premium SMS typically involves a text-messaging component, whereby a consumer purportedly authorizes charges by texting a particular five or six-digit number known as a "short code." Since the adoption of smartphones with advanced mobile web browsing capabilities and the greater use of mobile apps, there has been an increasing use of other forms of third-party billing arrangements, known as "direct carrier billing" arrangements. In direct carrier billing arrangements, a consumer does not necessarily need to send or receive a text message to initiate or complete a transaction that is billed to a mobile account. Instead, a consumer can initiate a transaction on a mobile website or within a mobile app, and the merchant can have the charge placed on the consumer's mobile account through back-end arrangements that involve the mobile carriers. In late 2013, after the Commission had held its mobile cramming roundtable and federal and state agencies had brought numerous law enforcement actions highlighting the prevalence of mobile cramming, the four largest mobile carriers stated their intention to discontinue one form of third-party billing – Premium SMS billing for commercial transactions.[38] Direct carrier billing, in contrast, is expected to continue growing, and it appears likely to supplant Premium SMS as the preferred mode of carrier billing. Regardless of the type of carrier billing involved, it is important for companies to provide basic consumer protections.

Providing Consumers the Option to Block Third-Party Charges

The Commission has advocated that mobile providers give consumers the option to block all third-party charges from their mobile phone accounts.[39] Providing a blocking option would significantly benefit consumers who wish to avoid third-party charges while imposing minimal costs to consumers who wish to use their mobile accounts for third-party billing. At activation, consumers should be informed that third-party charges may be placed on their accounts, and they should be given the opportunity to block all charges at that time. This option should be clearly and prominently disclosed to consumers while the accounts are active, including on the carriers' websites.

Staff also suggests that carriers should consider offering consumers the ability to block or allow only specific providers, or to block commercial providers only, as this may benefit consumers who wish to use their mobile accounts for only certain kinds of third-party charges. Allowing more granular blocking would permit consumers to continue to authorize third-party charges such as charitable or political donations.[40]

Strategies for Detecting and Preventing Mobile Cramming

Industry participants have adopted a range of strategies to attempt to detect and prevent mobile cramming. The staff report discusses many of these in detail and recommends best practices for improvement. These strategies address two key issues: avoiding deceptive practices that lead to unauthorized charges on mobile accounts, and ensuring that consumers are providing express, informed consent to third-party charges on mobile accounts.

The staff report notes that merchants are responsible in the first instance for ensuring that their practices – including any advertising, marketing, and opt-in processes – are not deceptive, pursuant to the FTC Act. Further, information about price is important to consumers and should be disclosed clearly and conspicuously before charging a consumer's telephone account for a good or service.[41] Thus, at a minimum, pricing information should be on the same page and immediately next to the purchase or buy button, entry of a PIN, or other invitation for a consumer to agree to a charge for a product or service. Additionally, advertising and purchase confirmation screens should clearly disclose that the charge is being billed to a specific telephone account. While industry guidelines have in the past focused extensively on the text-message based Premium SMS opt-in process, the basic consumer protection principles outlined in the report should apply regardless of the type of carrier billing used.

The staff report also recommends that carriers and billing intermediaries should implement reasonable procedures to scrutinize risky or suspicious merchants and terminate or take other appropriate steps against companies engaging in unlawful practices. For example, the report recommends that if a carrier or billing intermediary discovers that a merchant has run a campaign containing deceptive advertising, or discovers the merchant engaged in unauthorized billing on landline phones, the carrier or intermediary should closely monitor other campaigns run by that third party or its affiliates to ensure compliance.[42] Carriers and intermediaries can use monitoring techniques that compensate for known tactics that fraudsters use to evade detection of deceptive advertisements and sign-up processes. Industry participants also can adopt a policy of terminating serious and repeat offenders.[43]

Additionally, the report recommends that industry take stronger steps to ensure that consumers have opted-in to charges as represented by merchants. In Premium SMS, mobile carriers typically have relied on the merchant's representation – passed on by the billing intermediary – that a consumer opted-in to a charge. However, as the enforcement actions described above demonstrate, those representations are often unreliable. One option is to move toward more centralized control of the consumer opt-in process and authorization records, which appears to be the trend for at least some part of the industry.[44]

Finally, the staff report notes that monitoring consumer refund requests, and taking appropriate action when there are indications of unauthorized charges, can be a highly effective means of detecting and stopping cramming. Businesses providing other payment mechanisms use similar approaches to root out unauthorized charges. For example, credit card networks typically investigate merchants with chargeback rates of 1%, a threshold that is less than one-tenth of the refund rates seen in the cramming context.[45] While refund rates may differ across different types of payment methods, a representative from the Mobile Giving Foundation has suggested that charitable donations charged to a mobile bill and processed through the Foundation typically have a refund rate of under 1% overall.[46]

Adequate Disclosure of Third-Party Charges

Another important step in preventing cramming is ensuring that consumers are adequately informed of all third-party charges on their accounts. Carriers should clearly and conspicuously disclose all charges for third-party services in a non-deceptive manner. In particular, the name of the third-party service and any associated bill heading should relate to the product offered and not suggest an affiliation with the carrier's service. And, in order for carriers to make these disclosures, billing intermediaries and merchants should provide accurate information about these charges to them.

For consumers who auto-pay their bills, and may be especially unlikely to review the charges, or consumers who have prepaid phone plans, staff has urged carriers to consider whether a separate notification of third-party charges is warranted.

Consumer Dispute Protections and Refunds

The Commission has explained that mobile carriers should provide a clear and consistent process for customers to dispute suspicious charges on their mobile accounts and obtain reimbursement.[47] And indeed, FTC enforcement actions show that it is difficult for consumers to obtain refunds, and that refunds often are limited to only some months' worth of charges, even when consumers discover they incurred crammed charges for a longer time period.[48] A clear and consistent process is particularly important in this context because no federal statutory protections have been applied to consumer disputes about unauthorized charges placed on mobile carrier accounts. Consumers therefore have different dispute rights when using carrier billing than when using other payment mechanisms. For example, consumers have dispute resolution rights and liability limits for unauthorized credit card charges under Regulation Z, including a right to withhold payment while the dispute is pending, [49] and for unauthorized debit card charges under Regulation E, including a requirement that funds debited in an unauthorized transaction be returned to a consumer's account within ten days, pending further investigation.[50]

The staff report further suggests that mobile carriers also can do more to provide redress to consumers who have been crammed. For example, in the landline billing context, industry members have stated that consumers can withhold payment on disputed charges during the dispute period without a cut-off in phone service or accrual of interest. Industry should extend this protection to the mobile billing context, and inform consumers about it. The staff report also suggests that carriers be more proactive in notifying consumers when a third party's billing activities are terminated for unauthorized charges, in order to allow them to request a refund if appropriate.

CONCLUSION

Thank you for the opportunity to provide the Commission's views on mobile cramming. The Commission is committed to protecting consumers from mobile cramming and we look forward to continuing to work with the Committee and Congress on this important issue.

End Notes

[1] While the views expressed in this statement represent the view of the Commission, my oral presentation and responses to questions are my own and do not necessarily reflect the view of the Commission or any Commissioner.

[2] The FTC has jurisdiction under the FTC Act over market participants engaged in third-party billing. See FTC v. Verity Int'l, Ltd., 443 F.3d 48, 59-60 (2d Cir. 2006); In re Detariffing of Billing and Collection Servs., 102 F.C.C.2d 1150 30-34 (1986).

[3] To date, defendants have stipulated to final judgments, partially suspended based on inability to pay, totaling more than $160 million. See FTC v. Wise Media, LLC, No. 1:13-cv-1234-WSD (N.D. Ga. 2013); FTC v. Jesta Digital, LLC, No. 1:13-cv-01272 (D.D.C. 2103); FTC v. Tatto, Inc., No. 2:13-cv-08912- DSF-FFM (C.D. Cal. 2013). See also FTC v. Acquinity Interactive, LLC, No. 14-60166-CIV (S.D. Fla.) (amended complaint filed June 16, 2014); FTC v. MDK Media, Inc., No. 2:14-cv-05099-JFW-SH (C.D. Cal.) (complaint filed July 3, 2014).

[4] FTC v. T-Mobile USA, Inc., No. 2:14-cv-00967 (W.D. Wash. filed July 1, 2014).

[5] See Reply Comment of the Federal Trade Comm'n, Fed. Commc'ns Comm'n CG Docket No. 11-116 (July 20, 2012), at 7, 12, available at http://www.ftc.gov/sites/default/files/documents/advocacy documents/ftc-reply-comment-federal-communications-commission-concerning-placement-unauthorized-charges/120723crammingcomment.pdf [hereinafter "FTC Reply Comment"].

[6] See Fed. Trade Comm'n, Mobile Cramming Roundtable (May 8, 2013), available at http://www.ftc.gov/news-events/events-calendar/2013/05/mobile-cramming-ftc-roundtable [hereinafter "Roundtable"].

[7] See Fed. Trade Comm'n Staff, Mobile Cramming: An FTC Staff Report (2014), available at http://www.ftc.gov/system/files/documents/reports/mobile-cramming-federal-trade-commission-staff-report-july2014/140728mobilecramming.pdf [hereinafter "Mobile Cramming Report"].

[8] See MAJORITY STAFF OF S. COMM. ON COMMERCE, SCI., & TRANSP., OFFICE OF OVERSIGHT & INVESTIGATIONS, UNAUTHORIZED CHARGES ON TELEPHONE

BILLS, (July 12, 2011), at ii, available at http://www.commerce.senate.gov/ public/?a=Files.Serve&File id=3295866e-d4ba-4297-bd26- 571665f40756.

[9] See, e.g., FTC v. Hold Billing Servs., Ltd., No. 98-cv-00629-FB (W.D. Tex.) (contempt motion filed Mar. 28, 2012); FTC v. Inc21.com Corp., 745 F. Supp. 2d 975 (N.D. Cal. 2010), aff'd, 2012 WL 1065543 (9th Cir. Mar. 30, 2012); FTC v. Nationwide Connections, Inc., No. 06-80180 (S.D. Fla. Sept. 18, 2008) (stipulated order).

[10] Fed. Trade Comm'n, Examining Phone Bill Cramming: A Discussion (May 11, 2011), available at http://www.ftc.gov/bcp/workshops/cramming.

[11] See Comment of the Fed. Trade Comm'n, Fed. Commc'ns Comm'n CG Docket No. 11-116 (Oct. 24, 2011), at 5-6, available at http://www.ftc.gov/os/2011/12/111227cramming comment.pdf.

[12] See, e.g., Cecilia Kang, A Nation Outnumbered By Gadgets, Washington Post, Oct. 12, 2011, available at http://www.washingtonpost.com/business/economy/a-nation-outnumbered-by-gadgets/2011/10/11/gIQAhjdhdL story.html.

[13] See FTC Reply Comment, supra note 5.

[14] See Roundtable, supra note 6.

[15] See FED. TRADE COMM'N STAFF, PAPER, PLASTIC...OR MOBILE? AN FTC WORKSHOP ON MOBILE PAYMENTS (2013), at 7-8, available at http://www.ftc.gov/sites/default/files/documents/reports/paperplastic-or-mobile-ftc-workshop-mobile-payments/p0124908 mobile payments workshop report 02-28- 13.pdf.

[16] Mobile Cramming Report, supra note 7.

[17] Complaint for Permanent Injunction and Other Equitable Relief, at 10, FTC v. Tatto, Inc., No. 2:13-cv08912-DSF-FFM (C.D. Cal. Dec. 5, 2013), available at http://www.ftc.gov/ sites/default/files/documents/cases/131216bullroarercmpt.pdf.

[18] See id.; Memorandum In Support of Plaintiff's Ex Parte Application For Temporary Restraining Order With An Asset Freeze and Other Equitable Relief, And Order to Show Cause Why A Preliminary Injunction Should Not Issue, at 12, FTC v. Tatto, Inc., No. 2:13-cv-08912-DSF-FFM (C.D. Cal. Dec. 5, 2013).

[19] See Stipulated Order for Permanent Injunction and Monetary Judgment Against Defendants Tatto, Inc., Shaboom Media, LLC, Bune, LLC, Mobile Media Products, LLC, Chairman Ventures, LLC, Galactic Media, LLC, Virtus Media, LLC, and Lin Miao, FTC v. Tatto, Inc., No. 2:13-cv-08912-DSF-FFM (C.D. Cal. June 11, 2014), available at http://www.ftc.gov/system/files/documents/cases/140613bullroarerstiporder.pdf. The judgment was partially suspended based on defendants' inability to pay, but the defendants that have settled to date have surrendered more than $10 million in assets to be used for restitution.

[20] Complaint for Permanent Injunction and Other Equitable Relief, at 8-25, FTC v. Jesta Digital, LLC, No. 1:13-cv-01272 (D.D.C. Aug. 20, 2013), available at http://www.ftc.gov/ sites/default/files/documents/cases/2013/08/130821jestacmpt.pdf [hereinafter "Jesta Digital Complaint"].

[21] WAP opt-in involves consumers responding to an offer displayed on the mobile web by clicking on a confirmation button from the phone two separate times. This process captures the consumer's phone number without the need for the consumer to enter it manually.

[22] See Jesta Digital Complaint, supra note 20, at 8-28.

[23] See Stipulated Final Order for Permanent Injunction and Monetary Judgment Against Jesta Digital, LLC, FTC v. Jesta Digital, LLC, No. 1:13-cv-01372 (D.D.C. Aug. 23, 2013).

[24] Complaint for Permanent Injunction and Other Equitable Relief, at 7-8, FTC v. Wise Media, LLC, No. 1:13-cv-1234-WSD (N.D. Ga. Apr. 16, 2013), available at http://www.ftc.gov/sites/default/files/documents/cases/2013/04/130417wisemediacmpt.pdf.

[25] Memorandum in Support of Motion for Temporary Restraining Order, at 6, 10-11, FTC v. Wise Media, LLC, No. 1:13-cv-1234-WSD (N.D. Ga. Apr. 16, 2013) [hereinafter "Wise Media TRO Memo"].

[26] Id. at 6-9.

[27] Stipulated Order for Permanent Injunction and Monetary Judgment Against Defendants Brian M. Buckley and Wise Media, LLC, at 4-6, FTC v. Wise Media, LLC, No. 1:13-cv-1234-WSD (N.D. Ga. Nov. 22, 2013), available at http://www.ftc.gov/sites/default/files/documents/cases/131121wisemediabuckleystip.pdf.

[28] Amended Complaint for Permanent Injunction and Other Equitable Relief, FTC v. Acquinity Interactive, LLC, No. 14-60166-CIV (S.D. Fla. June 16, 2014), available at http://www.ftc.gov/system/files/documents/cases/140707revenuepathcmpt.pdf.

[29] Complaint for Permanent Injunction and Other Equitable Relief, FTC v. MDK Media, Inc., No. 2:14- cv-05099-JFW-SH (C.D. Cal. July 3, 2014).

[30] See Press Release, FCC Investigates Cramming Complaints Against T-Mobile (July 1, 2014), available at http://www.fcc.gov/document/fcc-investigates-cramming-complaints-against-t-mobile.

[31] See Complaint for Permanent Injunction and Other Equitable Relief, FTC v. T-Mobile USA, Inc., No. 2:14-cv-00967, 11-20 (W.D. Wash. July 1, 2014) [hereinafter "T-Mobile Complaint"]

[32] See id., 21-36.

[33] State law enforcement actions are discussed in more detail at pages 11-12 of the Mobile Cramming Report, supra note 7. The fact patterns described by the states are similar to those described in the Commission's actions.

[34] See Mobile Cramming Report, supra note 7, at 14-15, 17-18.

[35] Id. at 14, 33.

[36] See Wise Media TRO Memo, supra note 25, at 11-12; Mobile Cramming Report, supra note 7, at 14.

[37] Until recently, the Mobile Marketing Association ("MMA"), a trade association that promotes mobile marketing, had taken the lead in publishing best practices for merchants who wish to place charges on mobile phone bills using Premium SMS, but the MMA itself did not enforce those best practices. See Mobile Cramming Report, supra note 7, at 23-25.

[38] See, e.g., Ina Fried, AT&T, Sprint, T-Mobile, Verizon Dropping Most Premium Test Service Billing in Effort to Combat Fraud, ALLTHINGSD.COM, Nov. 21, 2013, http://allthingsd.com/20131121/att-sprint-tmobile-verizon-all-dropping-most-premium-text-service-billing-in-effort-to-combat-fraud/.

[39] See FTC Reply Comment, supra note 5, at 12.

[40] Mobile Cramming Report, supra note 7, at 22.

[41] See FED. TRADE COMM'N STAFF, REVISED .COM DISCLOSURES: HOW TO MAKE EFFECTIVE DISCLOSURES IN DIGITAL ADVERTISING (2013), at 10, available at http://www.ftc.gov/sites/default/files/attachments/press-releases/ftc-staff-revises-online-advertisingdisclosure-guidelines/130312dotcomdisclosures.pdf.

[42] Mobile Cramming Report, supra note 7, at 26-27.

[43] While there are costs to effective monitoring, there are also substantial benefits to both industry and to consumers. Industry participants can lower expenses related to the

processing of refund requests and handling of customer complaints. And consumers avoid being crammed with unauthorized charges.

[44] See Mobile Cramming Report, supra note 7, at 28-30. Centralization may shift some compliance costs, in the short term, from the merchants to carriers and billing intermediaries. However, it should benefit consumers and industry participants by making it more difficult for unscrupulous merchants to place unauthorized charges and by streamlining dispute resolution when a consumer claims a charge was unauthorized.

[45] See id. at 13-14 For example, in the Wise Media case, the monthly refund rates for some services on one carrier were as high as 40%. Wise Media TRO Memo, supra note 24, at 10.

[46] See Fed. Trade Comm'n, Mobile Cramming Roundtable Transcript (May 8, 2013), J. Manis, Mobile Giving Foundation, at 58, available at www.ftc.gov/sites/default/files/documents/ public events/Mobile%20Cramming%20Roundtable/30508m ob.pdf.

[47] FTC Reply Comment, supra note 5, at 12.

[48] Mobile Cramming Report, supra note 7, at 14, 33.

[49] See 12 C.F.R. §§ 1026.12, 1026.13.

[50] See 12 C.F.R. §§ 1005.6, 1005.11.

In: Cramming and Third Party Billing … ISBN: 978-1-63463-121-1
Editor: Christopher Rye

Chapter 5

TESTIMONY OF WILLIAM H. SORRELL, ATTORNEY GENERAL, STATE OF VERMONT. HEARING ON "CRAMMING ON WIRELESS PHONE BILLS: A REVIEW OF CONSUMER PROTECTION PRACTICES AND GAPS"*

SUMMARY

Chairman Rockefeller, Ranking Member Thune, and members of the Committee, thank you for the opportunity to testify today. The placement of unauthorized charges on telephone bills, also known as "cramming," has victimized many of my constituents in Vermont, and many of your constituents as well, including consumers, small businesses, and even large organizations. Cramming is a practice of significant interest to me, and one that the Vermont Attorney General's Office has been combatting, on behalf of Vermonters and citizens nationwide, for well over a decade.

Cramming is a huge, nationwide problem that has been pervasive in both landline and mobile telecommunications and has cost American consumers many billions of dollars. Cramming involves consumers being charged amounts on their phone bills without authorization, usually for goods and services "sold" by third-party vendors ranging from $9.99 to $24.99[1] per month that the consumer neither requested nor used. Among the things that make cramming so pernicious and persistent is the continuing lack of consumer awareness that their trusted telephone

* This is an edited, reformatted and augmented version of testimony presented July 30, 2014 before the Senate Committee on Commerce, Science, and Transportation.

carriers are able and willing to place charges on their telephone bills for goods and services sold by disreputable third parties. Not only do consumers not expect unanticipated third-party charges on their bill, they rarely recognize the charges that do appear on their bills as unauthorized third-party charges. Those few consumers that do detect unauthorized third-party charges—at least with respect to mobile cramming—have not consistently been able to obtain full refunds from their carriers or the carriers' third-party partners.

A number of regulatory approaches have proven to be ineffective in curbing cramming. In both the landline and mobile contexts, the telecommunications industry has largely been permitted to engage in "self-regulation." As this laissez-faire approach evinced its failure with respect to landline cramming, Vermont tried a notice-regime, requiring consumers to be notified in writing before receiving a third-party charge from their landline carrier. It is under these failed policies that cramming blossomed into the national, industry-wide, multi-billion dollar problem law enforcement officials and regulators—and, increasingly, consumers—across the country are familiar with today.

CRAMMING IS A HUGE PROBLEM THAT HAS COST CONSUMERS MANY BILLIONS OF DOLLARS

As the Committee is well aware, cramming has been recognized by many for its size and cost to consumers. In July 2011, this Committee concluded that landline cramming—a problem that had then been in existence for over a decade—had cost consumers a "substantial percentage" of $2 billion annually in "recent years."[2] In 2012, Consumer Reports estimated that landline and mobile cramming together could be costing American consumers up to $2 billion per year.[3]

Based upon Vermont's experience, if anything, these estimates of consumer loss are low. To date, as a result of my Office's investigations into dozens of third-party merchants and billing aggregators involved in *landline* cramming, 25,000 Vermonters have recouped nearly $2.3 million in crammed landline charges.[4] My Office has no reason to believe that these companies disproportionately targeted Vermont consumers. Moreover, there are many more such companies that perpetrated cramming that my Office has *not* investigated.[5] Thus, it is reasonable to conclude that approximately 1.25 million Americans have lost a staggering $1.15 *billion* to the entities Vermont has investigated and settled with to date *alone* and that these figures represent just a fraction of total consumer loss due to landline cramming.[6]

On the wireless side, the magnitude of consumer loss is equally daunting. As the Committee may be aware, my Office recently retained the University of Vermont's Center for Rural Studies to conduct a survey to determine the mobile cramming rate in Vermont for the customers of two major wireless carriers—that is, the proportion of third-party charges placed on mobile phones that were unauthorized.[7] Through the study, a sample of 2,400 Vermonters who had third-party charges placed on their mobile telephone bills during August and/or September of 2012 were contacted by my Office; we asked them about 5,388 third-party charges for a total of $43,250.96 that had been placed on their mobile phone bills over the course of those two months. Over 60% of the surveyed consumers reported the charges were crammed, bringing consumer losses over a two month period *for these 2,400 Vermonters alone* to over $25,950.58. Extrapolating nationwide, the Vermont survey suggests a similar survey done on a national scale would reveal that a sample of 1.2 million Americans lost $12,975,290 to mobile cramming during August and September of 2012 *alone*. My Office believes that carriers started placing third-party charges on mobile phone bills in Vermont as early as 2006. Even if only 5% of American consumers have ever experienced third-party charges on their mobile phone bills, consumer losses for mobile cramming alone may have exceeded $10 billion between 2007 and 2013.[8]

CRAMMING IS A PERVASIVE, NATIONWIDE AND INDUSTRYWIDE PROBLEM

This Committee is well aware of the depth and breadth of the landline cramming problem; I expect the Committee's work in recent years has uncovered the fact that mobile cramming is similarly nation and industrywide. Vermont, like other jurisdictions around the country, fielded an increasing number of consumer complaints about mobile cramming from 2006 to 2013.[9] While these complaints are voluminous, it is generally believed that they represent only a very small fraction of the consumers who have been improperly charged for third-party goods and services on their mobile phone bills.[10] Nevertheless, consumer complaints implicate the entire third-party charge industry; naming more than a dozen mobile carriers and hundreds of third parties, including content providers and billing aggregators.[11]

Importantly, mobile cramming is a problem that victimizes consumers no matter what mobile carrier they choose. Indeed, in a recent study of over 750 consumer complaints received by 28 jurisdictions, several state attorneys general discovered that 14 mobile carriers were implicated in cramming. Moreover, the following themes of consumer complaints were consistent across those carriers as well as across the country, and across time:

> Typically, consumers complain of having been signed up for a premium text messaging subscription service (or "PSMS" subscription) without their knowledge or authorization, costing them $9.99 or more on their mobile phone bill each month.
>
> Some of these subscriptions purport to be for goods such as ringtones and wallpaper, but many more are for "alerts"—a service in which the content provider purports to send periodic texts to the consumer with information about weather, traffic, news, or sports, for example, or with inspirational messages, horoscopes, celebrity gossip, or trivia.
>
> Consumers typically complain that they do not desire and do not use the goods and services for which they are being billed. Some consumers report that they do not even receive the alerts for which they have been charged.
>
> While consumers can sometimes recall having gotten spam text(s) or having entered their mobile phone numbers into a website immediately prior to being signed up for a subscription service (often to receive a "free" good or service), just as often— if not more often—consumers simply have absolutely no idea how they came to be signed up for the subscription.
>
> Consumers are often crammed by more than one content provider and, many times, after they have already asked their mobile carrier to place a block on their account to stop third party billing altogether.
>
> Consumers that detect that they have been crammed on their mobile phone bills typically do so after they have been paying for a subscription service for several months.
>
> Even consumers who do not text and have no access to the internet (and thus, cannot have opted in to a third party good or service through a typical method) report having been crammed. Too often, these consumers are elderly.[12]

These are the very same issues my Office hears about when it communicates with Vermont consumers about mobile cramming.[13] Thus, it should come as no surprise that my Office believes that many of the allegations contained in the Federal Trade Commission's recent complaint against T-Mobile are representative of behaviors engaged in by carriers throughout the industry.[14]

MOST THIRD-PARTY CHARGES ARE CRAMMED CHARGES

It is my Office's conclusion that most third-party charges are crammed charges. My Office has conducted three consumer surveys regarding cramming since the Fall of 2011. According to these surveys, in excess of 60% of third party charges are unauthorized ("crammed"). As stated above, according to our 2013 mobile third party authorization study, over 60% of third-party charges on mobile phone bills were crammed charges. Previously, in the fall of 2011, my Office spoke to over 100 Vermont consumers by phone about their experience with third-party charges on their mobile telephone bills; a full 92% of the consumers said the charge was crammed, while only 8% of the consumers reported that the charge was authorized.[15] Finally, nearly 90% of consumers surveyed in connection with my Office's first landline billing aggregator investigation reported that the third-party charges on their landline bills were unauthorized.[16] We are aware of no studies that contradict these findings.

THIRD-PARTY CHARGES HAVE ENRICHED INDUSTRY WHILE OFFERING CONSUMERS LITTLE VALUE

It is no secret that the landline and mobile telecommunications industries have profited to the tune of billions of dollars from placing third-party charges on landline and mobile telephone bills.[17] And yet, consumers have gained very little as a result. Consumers not only report that they are crammed for third-party services that they did not want or authorize, but that they would have no reason to value such offerings. A number of the landline consumers my Office spoke to in connection with our first landline billing aggregator investigation indicated they had no reason to order the voicemail service for which they were charged; the respondents gave such explanations as, "[I] have an answering machine [and so] would never use this service," "I had voicemail from the phone company [and] did not need [another service]," and "[I] can't imagine agreeing to voicemail since we have always had a personal voice recorder."[18] Further, 73% of the consumers my Office interviewed in 2011 said they would have had no reason to purchase the goods or services for which they were billed on their mobile phone—for example, one consumer had been charged for stock alerts, but owned no stocks and did not follow the market.[19] Finally, while consumers typically complain they do not desire and do not use the goods and services for which they are charged on their phone bills, some consumers report they have not even *received* the goods and services for which they have been charged.[20] It is, therefore, of no surprise that there was no consumer outcry following the mobile carriers' decision to exit the Premium Short Message Service ("PSMS") platform—widely believed to be responsible for the lion's share of the mobile cramming problem—in November of 2013. Likewise, my Office has received no negative feedback from consumers following Vermont's passage of the 2011 ban on most third-party charges on landline bills. The only logical conclusion: contrary to industry talking points,[21] very few, if any, consumers received any real value from this billion-dollar industry.

CRAMMING GOES UNDETECTED BY CONSUMERS

Cramming is a particularly serious problem because consumers do not know the underlying activity—the placement of third-party charges on phone bills—is happening. Consumers do not expect third-party charges to appear on their phone bills because they do not understand it is possible for third parties to charge them this way. Often third-party charges appear on phone bills without the consumer having taken any action at all. This problem is exacerbated by the fact that such charges are not readily discernible on the billing statements.

The data consistently show that consumer awareness about the third-party charges on their landline and mobile telephone bills is very low in Vermont. According to our 2013 mobile cramming survey, in excess of 78% of the consumers reported that, prior to receiving the survey, they had been unaware they could be billed for goods and services provided by third parties on their mobile phone bills.[22] In my Office's 2011 mobile cramming survey, significantly more than half of the consumers did not know that the third-party charge was on their bill until they were informed of the charge by my Office, nor did they know that they could be billed for third-party goods and services on their mobile phone bills prior to being informed of the charge in question.[23] Finally, according to the consumers surveyed in connection with my Office's first landline aggregator investigation, only 27.4% of consumers noticed the third-party charge on their landline bills within three months of being charged.[24]

The national picture is no different; consumers often do not know how they came to be signed up for third-party charges on their phone bills. A recent national mobile-cramming complaint analysis indicated the following were among the themes of consumer complaints:

> While consumers can sometimes recall having gotten spam text(s) or having entered their mobile phone numbers into a website immediately prior to being signed up for a subscription service (often to receive a "free" good or service), just as often— if not more often—consumers simply have absolutely no idea how they came to be signed up for the subscription.
>
> Consumers that detect that they have been crammed on their mobile phone bills typically do so after they have been paying for a subscription service for several months.
>
> Even consumers who do not text and have no access to the internet (and thus, cannot have opted in to a third party good or service through a typical double opt-in method) report having been crammed. Too often, these consumers are elderly.[25]

Consumers also routinely report that they do not understand their mobile phone bills and/or they find it difficult to detect the source of the third-party charges appearing on their bills.[26] Some consumers have even complained of having been crammed by third parties whose names (or whose subscription/product names) make it very difficult to detect that a third party was involved.[27] Consumers often express confusion about their mobile phone bills, and report that they had no idea they could be charged on their mobile phone bills for goods and services provided by third parties.[28] As a result, consumers are typically charged on their mobile phone bills for third-party subscriptions for multiple months before they recognize they have been crammed.[29]

CONSUMERS DO NOT OBTAIN FULL REFUNDS FOR CRAMMING

At least with respect to mobile cramming, consumer experience with obtaining refunds is inconsistent. While some consumers are able to get full refunds from their mobile carrier, many are not.[30] Vermont consumers have likewise reported mixed success with obtaining refunds from their mobile carriers; while some are able to obtain full refunds, others are only able to obtain a partial refund.[31] Still others are unable to get any refund from their mobile carrier, or are promised refunds they never receive.[32] Consumers also report that carriers refer them to the third-party content provider to seek refunds, and/or that they are unable to reach the content provider, or, if they do reach the content provider, are unable to get a full, or even partial, refund.[33]

THE TIME IS RIPE FOR A NEW APPROACH

After over a decade of fighting landline cramming with enforcement actions against third-party merchants and billing aggregators, my Office successfully advocated for legislation prohibiting most third-party charges on landline telephone bills.[34] This statutory approach takes account of actual consumer expectations—*i.e.,* that consumers do *not* anticipate they will be charged on their phone bills for third-party goods and services—is straightforward to enforce, and does not interfere with other methods of receiving payment for services provided.[35] Most importantly, we believe the law has *worked.*[36]

My Office decided to take another approach when turning our attention to mobile cramming. After launching dozens of investigations into third-party content providers and mobile billing aggregators, my Office began pursuing *carriers* for billing for unauthorized charges, first on behalf of Vermont, and then on behalf of—now—46 states.[37] In November of 2013, the nation's four largest mobile carriers—Verizon, AT&T, Sprint and T-Mobile—decided to stop charging their customers for commercial PSMS, a platform which accounted for the majority of third-party charges on mobile phones, and for the overwhelming majority of cramming complaints.[38] My Office believes that billing for commercial PSMS has now ceased industry-wide, and has heard from Offices around the country that mobile-cramming complaints have slowed to a trickle, no doubt as a result. While the mobile carriers' exit of PSMS is undoubtedly very positive for consumers, it was a voluntary move on the part of industry and carries with it no guarantee of future action with regard to PSMS or other, similar platforms such as Direct to Consumer Billing ("DCB"). Non-PSMS mobile cramming complaints are few in number, but do exist.[39] Moreover, we expect platforms, such as DCB, that are more appropriate for consumption by consumers with smartphones (rather than feature phones, to which PSMS was keyed) to be on the rise.

It is my opinion that a new approach would be appropriate on a federal level as well. While my Office's legislative advocacy has effectively stopped landline cramming in Vermont, we believe our law-enforcement leadership has merely pressed the "pause" button on mobile cramming. As we look forward to a world that becomes more mobile-device oriented, the time is right for all of us to take stock of the major lessons learned—the potential for enormous consumer loss, low consumer awareness about the practice, the difficulty of getting consumer redress—to ensure that cramming does not fool us again.

End Notes

[1] These dollar amounts are typical for crammed charges on mobile phones.

[2] S. COMM. ON COMMERCE, SCI. AND TRANSP., 112TH CONG., UNAUTHORIZED CHARGES ON TELEPHONE BILLS: STAFF REPORT FOR CHAIRMAN ROCKEFELLER (July 12, 2011).

[3] Beware of Bogus Phone Bill Fees; Consumers Could Be Losing Up to $2 Billion a Year, CONSUMER REPORTS, August 2012, available at http://www.consumerreports.org/cro/magazine/2012/08/beware-ofbogus-phone-bill-fees.

[4] Press Release, Office of the Attorney General of Vermont, A.G. Settles Case for $1.6 Million in Ongoing Effort to Combat "Cramming" (Nov. 12, 2013), available at http://ago.vermont.gov/focus/news/a.g.-settles-case-for-$1.6-million-in-ongoing-effort-to-combat-cramming.php.

[5] Unauthorized Charges on Telephone Bills: Why Crammers Win and Consumers Lose: Hearing before the S. Comm. on Commerce, Sci., and Transp., 112th Cong. 112-171 (2011) [hereinafter Hearing] (statement of Elliot Burg, Senior Assistant Atty Gen., State of Vermont).

[6] See State & County QuickFacts, Vermont, UNITED STATES CENSUS BUREAU (July 8, 2014), http://quickfacts.census.gov/qfd/states/50000.html (estimating United States and Vermont populations for 2013).

[7] See Jane Kolodinsky, Mobile Phone Third-Party Charge Authorization Study: Vermont, CENTER FOR RURAL STUDIES AT THE UNIVERSITY OF VERMONT (May 5, 2013). available at http://ago.vermont.gov/assets/files/Mobile%20Phone%20ThirdParty%20Charge%20Authorization%20Study.pdf

[8] Based on the 84 months from January 2007 through December 2013; while my Office's understanding is that the practice of placement of third-party charges on mobile telephones began as early as 2004, we are not aware of cramming complaints predating 2006.

[9] Letter from Nat'l Ass'n of Att'ys Gen. to Donald Clark, Sec'y, Fed. Trade Comm'n 2-3 (June 24, 2013) (on file with Fed. Trade Comm'n), available at: http://www.ftc.gov/sites/default/files/documents/publiccomments/2013/06/564482-00015- 86106.pdf [hereinafter Letter].

[10] Id.

[11] Id.

[12] Letter, supra note 9, at 3 – 4.

[13] Id. at 4, 6.

[14] See Complaint, Fed. Trade Comm'n v. T-Mobile USA, Inc., No. 2:14-cv-00967 (W.D. Wash. July 1, 2014), available at http://www.ftc.gov/system/files/documents/cases/140701 tmobilecmpt.pdf (alleging, inter alia, that T-Mobile charged consumers for unauthorized third-party subscriptions despite clear indications that the charges were unauthorized, that

T-Mobile retained a significant portion of the revenue obtained through such charges, and that T-Mobile bills obscured the nature and source of such charges on consumers' mobile telephone bills).

[15] Letter, supra note 9, at 4.

[16] Hearing, supra note 5, at 2.

[17] S. COMM. ON COMMERCE, SCI. AND TRANSP., 112TH CONG., supra note 2, at iii; Complaint, Fed. Trade Comm'n v. T-Mobile USA, Inc., No. 2:14-cv-00967, at 3 – 4.

[18] Hearing, supra note 5, at 2.

[19] Letter, supra note 9, at 4.

[20] Id. at 3.

[21] Michael F. Altschul, Senior Vice President and Gen. Counsel, CTIA: The Wireless Association, Fed. Trade Comm'n Workshop: Bill Shock and Cramming (Apr. 17, 2014, 9:00 AM) http://www.fcc.gov/events/workshop-bill-shock-and-cramming.

[22] Kolodinsky, supra note 7, at 8.

[23] Letter, supra note 9, at 4.

[24] Hearing, supra note 5, at 2.

[25] Letter, supra note 9, at 3 – 4.

[26] Id. at 7.

[27] Id. at 3, n. 4 (explaining that national consumers named the following third parties and/or campaigns in their complaints: "General Texting," "General Texting Co.," "General Texting, LLC," "General Texting.com," "Premium Axcess,"[sic], "Premium Customer Care," "Premium SMS," "Premium Text Messaging," "Text Savings," and "Text Savings, LLC.").

[28] Letter, supra note 9, at 7.

[29] Id.

[30] Id. at 3-4.

[31] Id. at 6.

[32] Id.

[33] Id. at 3-4.

[34] See 9 VT. STAT. ANN. tit. 9 § 2466 (2014) (prohibiting most third-party charges from being placed on Vermonters' landline telephone bills).

[35] Hearing, supra note 5, at 4.

[36] In the approximate two years since the ban became effective, my Office has received no more than 2 complaints about unauthorized third-party charges on landline bills that postdate the ban.

[37] See Press Release, Office of the Attorney General of Vermont, AT&T Mobility, Sprint and T-Mobile Will Stop Billing Problematic Third-Party Charges (November 21, 2013), available at http://ago.vermont.gov/focus/news/att-mobility-sprint-and-t-mobile-will-stop-billing-problematic-thirdparty-charges.php (referring to a then-45-state matter).

[38] Id.; Ina Fried, AT&T, Sprint, T-Mobile, Verizon Dropping Most Premium Text Service Billing in Effort to Combat Fraud, ALL THINGS D (November 21, 2013), http://allthingsd.com/20131121/att-sprint-tmobile-verizon-all-dropping-most-premium-text-service-billing-in-effort-to-combat-fraud/.

[39] Note also that my Office's 2013 study was of all third-party charges, and not just PSMS charges.

In: Cramming and Third Party Billing …
Editor: Christopher Rye
ISBN: 978-1-63463-121-1

Chapter 6

TESTIMONY OF TRAVIS LEBLANC, ACTING CHIEF, ENFORCEMENT BUREAU, FEDERAL COMMUNICATIONS COMMISSION. HEARING ON "CRAMMING ON WIRELESS PHONE BILLS: A REVIEW OF CONSUMER PROTECTION PRACTICES AND GAPS"*

Chairman Rockefeller, Ranking Member Thune, and members of the Committee, I am Travis LeBlanc, Acting Chief of the Enforcement Bureau at the Federal Communications Commission. Thank you for the opportunity to appear before you to highlight the FCC's efforts to combat the harmful practice of placing unexpected or unauthorized charges on consumers' telephone bills, a practice known as cramming.

THE CRAMMING PROBLEM

Cramming is a significant problem, and one that, by its nature, has caused countless consumers to unwittingly open their wallets for products and services they never wanted. A report released in 2011 by Chairman Rockefeller after a year-long investigation of cramming on landline telephone bills showed that telephone companies placed approximately $2 billion worth

* This is an edited, reformatted and augmented version of testimony presented July 30, 2014 before the Senate Committee on Commerce, Science, and Transportation.

of third-party charges on their subscribers' bills each year, and that most of these charges were unauthorized.[1] Fifteen to twenty million U.S. households are estimated to have been victims of cramming on their landline telephone bills,[2] and most do not even know it. Historically, many consumers have been completely unaware that their carriers are permitted to charge them for third-party products and services, and do not know to look for third-party charges on their telephone bills. Even those who are aware of the possibility often fail to spot unauthorized charges on their bills, because the charges have been hidden from scrutiny: they have been buried in multi-page bills, not clearly described, or small enough in amount to go unnoticed.[3] Consumer deception is a hallmark of cramming. The Commission took action in 2012 to help wireline consumers detect – or simply avoid – cramming, but unfortunately consumers continue to be crammed.

Today's hearing is about the fact that cramming is not just a problem for those with landline telephones. Since 2010, the FCC has received more than 5,000 complaints about cramming, and the proportion of those about unauthorized charges on wireless bills has grown from about 15% in 2008-2010 to 58% in 2013. Because so many consumers do not even realize that they have been crammed, or lack the time or knowledge to complain to the FCC, these numbers represent just the tip of the iceberg. A 2012 analysis by the Illinois Citizens Utility Board found that the percentage of fraudulent third-party charges on Illinois consumers' wireless bills skyrocketed in just one year, from about 26% to 51%.[4] It is critical that cramming on wireless bills not be overlooked, especially now, when Americans are becoming increasingly reliant on their mobile phones. The Pew Research Center estimates that 90% of American adults have a cell phone, including 74% of Americans 65 and over.[5] According to the Centers for Disease Control and Prevention, two in five U.S. households have "cut the cord" entirely from their landline phones and are using only mobile phones.[6]

Crammers are predators. They evolve with consumers. As consumers migrate to wireless phones and away from landlines, we expect that the same kind of predators that profited from unauthorized landline charges will look to wireless bills for new and creative ways to defraud consumers. Today, a majority of Americans (58%) have a smartphone.[7] The rise in wireless phone dependence introduces new ways for bad actors to profit from sneaky billing practices. This is because modern smartphones are not just phones that facilitate only voice communications, but sophisticated handheld computers that enable consumers to engage in a wide array of activities, from interactive gaming, to buying a coffee in a café, to shopping online from wherever they are. Consumers with Android phones, for example, can charge all of their app purchases to their phone bills, a form of direct-carrier billing.[8] The more consumers' mobile phone bills become like credit card bills – reflecting a host of different purchases – the more difficult it may become to spot unauthorized charges.

If there is any good news, it is that in 2012, major landline carriers announced that they would discontinue most third-party billing,[9] and in 2013, major wireless carriers followed suit,[10] at least with respect to third-party billing via premium short messaging services, or PSMS. Perhaps as a result of these agreements and increased government scrutiny, cramming complaints are trending downward. Of course, to the extent that the practice of cramming decreases due to private sector commitments, that is commendable. Unfortunately, that has not proven to be a silver bullet to the heart of cramming; we have not seen the practice of cramming cease entirely. Therefore, strong enforcement is still needed and perhaps additional regulation as well. Protecting consumers is the common goal that we all share, and the FCC stands ready to use the full spectrum of its authority to thwart bad actors and prevent consumers from being defrauded.

The FCC's Role in Combatting Cramming

As the federal agency with primary oversight of the nation's telephone carriers, the FCC approaches the problem of cramming through a combination of enforcement, regulation, and consumer education. On the enforcement side, the Commission has taken fourteen enforcement actions since 2010 for placing unauthorized charges on consumers' phone bills. Collectively, these actions are valued at no less than $122,750,000, including monetary forfeitures the Commission has sought to impose, payments the FCC has ordered enforcement targets to make to the U.S. Treasury in connection with settlements, and refunds the FCC has ordered the targets to make to injured consumers. Just in the last two weeks, the FCC has taken three of these enforcement actions, which proposed over $10.5 million in penalties and payments to the U.S. Treasury.

On the regulatory side, the FCC has adopted "truth-in-billing" rules designed to help consumers detect cramming or other unauthorized activities associated with their phone service, and is considering expansion of the rules. It is expected that the Commission will consider any rule changes within the next several months. We also anticipate that the Commission will conduct a workshop or similar event in light of continually evolving third-party billing technologies and practices. And on the education side, the FCC has been engaging consumers through written and video guides, tip sheets, and other materials aimed at empowering them to identify and report cramming.

The FCC's power to address cramming comes from its statutory authority over carriers. The Communications Act of 1934 is the FCC's enabling statute, and Section 201(b) is one of its cornerstones. That section declares unlawful all "unjust and unreasonable" charges and practices "for and in connection with" an "interstate or foreign or communication service by wire or radio."[11] Over fifteen years ago, the FCC found that cramming constituted an unjust and unreasonable practice.[12] Section 201(b), as well as Section 258, the anti-slamming provision of the Act, are the sources of authority for its truth-in-billing rules.

ENFORCEMENT

Under the Communications Act, the FCC has a variety of enforcement tools available to achieve compliance. Most often, the FCC initiates a forfeiture proceeding for violations of the Communications Act, including Section 201(b). Generally speaking, the first step in the process is for the FCC to issue a notice of apparent liability for forfeiture, or NAL. The Communications Act authorizes the FCC to impose a penalty of up to $160,000 for each violation, or each day of a continuing violation, up to a maximum of $1,575,000 for a continuing violation. Typically, the FCC proposes a forfeiture of $40,000 for each apparent cramming violation, although in recent cases it has substantially increased that amount for egregious violations. The Enforcement Bureau is generally open to settling a matter in lieu of initiating a forfeiture proceeding. As a condition of settlement, the FCC may require a carrier to reimburse consumers who were injured by its unlawful practices.

The FCC's cramming enforcement actions have arisen from three basic kinds of bad conduct. The first is what I will call "billing carrier cramming," and involves a carrier, either landline or wireless, billing its own subscriber for a third-party product or service. In connection with mobile service in the United States, this has most often involved the carrier charging its subscribers, via PSMS, around $10 per month for services such as flirting tips, horoscopes, lottery results, and stock quotes. This is the alleged fraudulent activity at issue in the FCC's recently-announced investigation of T-Mobile, and the Federal

Trade Commission's complaint in federal district court against the carrier.[13] The FCC and the Federal Trade Commission worked collaboratively on this investigation in order to harmonize our respective enforcement as well as to leverage our respective expertise. T-Mobile allegedly crammed hundreds of millions of dollars of PSMS charges onto its subscribers' phone bills from third parties whom the carrier knew, or should have known, did not have authorization to bill its subscribers. Indeed, some of the third parties had refund rates, or "charge-backs," of 40%, and some had been sued for fraud. We are pleased with our collaboration with the Federal Trade Commission on the T-Mobile investigation and look forward to continuing to partner with the Federal Trade Commission in the future.

The second type of FCC cramming enforcement action is what I will call "third-party carrier cramming." It involves a fraudulent carrier placing an unauthorized charge for its own product or service on a consumer's phone bill issued by another carrier, typically the consumer's local phone bill. Fraudulent conduct of this type was at issue, in four NALs the Commission released in 2011, which collectively proposed forfeitures of nearly $12 million.[14] These third-party carriers assessed charges for their own "dial-around" long-distance service of around $10-15 per month on consumers' local phone bills. Each of the third-party carriers assessed its charges on at least tens of thousands – if not hundreds of thousands – of bills for the service in the year preceding the enforcement action. When the FCC investigated how many consumers the carriers had actually provided service to during that time, incredibly, two of the carriers disclosed that they had serviced only about 20 to 25 consumers, and the other two carriers could or would not answer the question. Further, earlier this month, the FCC settled with another carrier, Assist 123, LLC, for allegedly placing unauthorized PSMS charges for subscription services like directory assistance, movie listings, driving directions, and lottery results, on consumers' landline and wireless phone bills. The settlement requires Assist 123 to pay $1.3 million to the U.S. Treasury.[15]

Another flavor of "third-party carrier cramming" is often connected with "slamming" – the unauthorized switch of a consumer's preferred carrier. In the last fourteen months, the FCC has issued five NALs against carriers for apparently switching or attempting to switch consumers' long-distance service through deceit and trickery, and then charging the consumers for a new carrier's service they did not authorize or want. Collectively, the NALs proposed forfeitures of nearly $28 million.[16] According to the NALs, the *modus operandi* for most of these third-party carriers involved their agents cold-calling consumers, pretending to be affiliated with a consumer's existing provider, offering improved or upgraded service, recording the consumer supposedly authorizing such improved or upgraded service with the existing provider, and then using that "authorization" to switch, or attempt to switch, the consumer's existing long-distance service to the third-party carrier. In several of these cases, the FCC found the conduct so egregious that it proposed additional penalties beyond doubling or tripling the "base" forfeiture of $40,000 for cramming. Indeed, in one enforcement action that the FCC took this month, the agency found that the carrier, Optic Internet Protocol, Inc., apparently not only may have tricked consumers to obtain purported authorization, but also fabricated the recordings it offered to regulatory authorities as proof of authorization.

In some of the slamming/cramming cases, the FCC has alleged that the rogue carrier also billed consumers directly for its service, because the rogue carrier was not successful in pushing its charges onto the consumers' local exchange carrier bill. Conduct of this type does not involve cramming for unauthorized third-party charges, because the carrier is placing the charge for its own service on its own bill. This conduct nevertheless involves placing an unauthorized charge on a telephone bill, and the FCC has found that it also constitutes an unjust and unreasonable practice that violates Section 201(b) of the Communications Act.[17]

Since 2010, the FCC has pursued two major wireless carriers for allegedly placing unauthorized charges for their own services on their own bills. Both cases involved the carriers allegedly charging their subscribers for data services the subscribers did not expressly authorize.[18] The FCC ordered both carriers to notify affected customers of the applicable charges, to offer refunds, and to make a payment to the U.S. Treasury in lieu of a penalty. One of the cases required the carrier to refund at least $52.8 million to affected consumers, and to pay $25 million to the U.S. Treasury.

Rulemaking/Regulation

The FCC is also addressing cramming on the regulatory side. Fifteen years ago, the FCC adopted its first "truth-in-billing" rules, in order to help consumers detect cramming, slamming, and other fraud in connection with their telephone bills and telecommunications services. In 2009, the FCC issued a Notice of Inquiry to explore whether, and if so, how, to amend its "truthin-billing" rules; in 2011, in response to continued cramming problems, the FCC issued a Notice of Proposed Rulemaking to strengthen its rules; and in 2012, the agency in fact did so.[19] Among other things, the current "truth-in-billing" rules require both landline and wireless carriers to clearly and conspicuously: (1) identify the name of each service provider associated with a billed charge; (2) identify any change in any service provider from the preceding billing cycle;

1) provide a brief, non-misleading, plain language description of the services billed; and
2) display a toll-free number for subscribers to dispute, or inquire about, any billed charge. The "truth-in-billing" rules also require landline – but not wireless – carriers to: (1) separate charges by service provider; (2) set forth charges from third parties for non-telecommunications services in a distinct section of the bill; and (3) notify subscribers that the carrier offers subscribers the opportunity to block third-party charges on their bills, if in fact the carrier does so.[20]

When the FCC strengthened its rules in 2012, it also issued a Further Notice of Proposed Rulemaking on whether it should expand the coverage of the rules still more. The FCC asked whether it should expand all of the existing rules to wireless carriers; whether it should prohibit carriers from assessing charges for third-party products and services on a subscriber's bill absent the subscriber expressly opting in; or whether it should altogether ban carriers from assessing charges for third-party products and services, at least on the same bills that contain charges for regulated telecommunications services. The FCC asked for additional comment on these and other issues last year, with the comment period officially closing in December 2013. The FCC is now poised for action in that docket, and expects to consider any rule changes within the next several months.

To keep abreast of new and emerging kinds of cramming as well as new carrier billing technologies and practices, the FCC is also planning to host a workshop or similar event on these topics in the next six months.

CONSUMER EDUCATION

In addition to its enforcement and regulatory work, the FCC also works to educate consumers about cramming. The agency has issued both printed and video consumer guides, as well as tip sheets on how to identify and report cramming.[21] The Commission also held a comprehensive public workshop on cramming in 2013. These outreach efforts, many of which involved close coordination with groups such as AARP, are intended to alert consumers to the fact that their carriers may charge them for third-party products and services on their telephone bills, and encourage consumers to review their bills carefully each month, paying attention to even small charges, as well as the descriptions offered for all charges, and who is responsible for them. The materials also encourage consumers to call their carriers about any charges they question, and any provider identified on the bill associated with such charges. In addition, our consumer education materials explain the FCC's "truth-in-billing" rules in plain English, and tell consumers how to file complaints with not only the FCC, but also the Federal Trade Commission and state public service commissions.

Enforcement Bureau Reforms

The FCC is proud of these strong enforcement actions, which are deterring cramming, but we will continue our commitment to do more to protect consumers. Since I joined the FCC four months ago, my first priority has been to make sure the great people and resources of the Enforcement Bureau are used as effectively and efficiently as possible. The Enforcement Bureau is embracing a modern enforcement philosophy that says we need to be smarter about how to deploy our limited resources. In many instances, that means working more closely and more frequently with our fellow law enforcement partners at the federal and state levels, just as we have in the cramming context.

We are focused on ensuring the widest possible compliance with the law and rules that have the most impact on Americans in the 21st Century. For example, America's growing reliance on wireless phones leaves them increasingly vulnerable to unlawful privacy-invading robocalls and text-message spam. Cellular and Global Positioning System (GPS) jammers that can disrupt critical infrastructure and public safety networks are more and more widely available and must also continue to be the focus our enforcement efforts. And the recently announced Strike Force within the Enforcement Bureau will combat fraud, waste, and abuse in the Universal Service Fund (USF). It is our duty to vigorously protect the integrity of the USF programs by ensuring that program funds are used for their intended purposes.

We are also in the vanguard of the FCC's Commission-wide Process Reform efforts. For many outside the Commission, our most significant efforts in this area are already evident in our firm commitment to speedy resolution of both routine and significant matters. The reforms we have already adopted, and those in process, will ensure that the Commission's team of prosecutors will have the best chance at doing the most good for the greatest number of Americans.

Fundamentally, we are creating an Enforcement Bureau that is an efficient and smart prosecutorial unit. We are striving to be data-driven, nimble, creative, strategic and collaborative. Ultimately, I hope this will be good for consumers, good for industry, and just plain good government.

CONCLUSION

The FCC is the nation's federal regulatory authority over telecommunications carriers, and Congress has empowered the agency to combat unjust and unreasonable practices by carriers, as well as to adopt rules governing the conduct of carriers. Through its enforcement, regulatory, and consumer education work, the FCC is actively using these powers to address cramming and other unlawful acts by carriers that place unexpected and unauthorized charges on consumers' phone bills. We look forward to continued cooperation with the Committee and other regulatory authorities at both the federal and state levels toward the common goal of protecting consumers from these unjust and unauthorized charges.

End Notes

[1] Unauthorized Charges on Telephone Bills, U.S. Senate Committee on Commerce, Science & Transportation, Office of Oversight & Investigations, Majority Staff, Staff Report for Chairman Rockefeller (rel. July 12, 2011), available at http://www.commerce.senate.gov /public/?a=Files.Serve&File id=3295866e-d4ba-4297-bd26-571665f40756.

[2] Empowering Consumers to Prevent and Detect Billing for Unauthorized Charges ("Cramming"), Report and Order and Further Notice of Proposed Rulemaking, 27 FCC Rcd 4436, 4437 (2012) ("FCC Cramming Order").

[3] Id. at 4444.

[4] Citizens Utility Board, Analysis: Frequency of Cellphone "Cramming" Scam Doubles in Illinois, CUB Concerned Wireless Customers Targeted as Landline Law Tighten (Dec. 4, 2012), available at http://www.citizensutilityboard/pdfs/NewsReleases/20121204 CellPhone Cramming.pdf.

[5] Pew Research Internet Project, Cell Phone and Smartphone Ownership Demographics, available at http://www.pewinternet.org/data-trend/mobile/cell-phone-and-smartphone-ownership-demographics/.

[6] Centers for Disease Control and Prevention, National Center for Health Statistics, Wireless Substitution: Early Release of Estimates From the National Health Interview Survey, July–December 2013, (July 2014), available at http://www.cdc.gov/nchs/data/nhis/early release/wireless201407.pdf.

[7] Pew Research Internet Project, Cell Phone and Smartphone Ownership Demographics, available at http://www.pewinternet.org/data-trend/mobile/cell-phone-and-smartphone-ownership-demographics/.

[8] See, e.g., Ingrid Lunden, "Amazon's Carrier Billing Deal With Bango To Kick In This Year – Changes For The Appstore And Amazon Ahead?," TechCrunch, Mar. 20, 2013, available at http://techcrunch.com/2013/03/20/amazons-carrier-billing-deal-with-bango-to-kick-in-this-year-changes-for-theappstore-and-amazon-ahead/; Kevin Parrish, "Google Play Offers Option to Charge Purchases to Your Bill," Tom's Guide, May 3, 2012, available at http://www.tomsguide.com/us/Google-Play-Carrier-Billing-AT-T-AndroidSprint,news-15070.html .

[9] Press Release, U.S. Senate Committee on Commerce, Science, & Transportation, Another Major Phone Company Agrees to End Third-Party Billing on Consumer Phone Bills (Mar. 28, 2012), available at http://www.commerce.senate.gov/public/index.cfm?p= Press Releases&ContentRecordid=0245033e-6fe4-420d8ed3-cdb39ed6537f.

[10] Press Release, Office of the Vermont Attorney General, AT&T Mobility, Sprint, and T-Mobile Will Stop Billing Problematic Third-Party Charges (Nov. 21, 2013), available at http://www.atg.state.vt.us/news/att-mobility-sprintand-t-mobile-will-stop-billing-proble matic- third-party-charges.php.

[11] 47 U.S.C. § 201(b).

[12] Long Distance Direct Direct, Inc., Notice of Apparent Liability for Forfeiture, 14 FCC Rcd 314 (1998).

[13] Press Release, Federal Communications Commission, FCC Investigates Cramming Complaints Against T-Mobile, (rel. July 1, 2014), available at http://transition.fcc.gov/ DailyReleases/DailyBusiness/2014/db0701/DOC327998A1.pdf.

[14] Cheap2Dial Telephone, LLC, Notice of Apparent Liability for Forfeiture, 26 FCC Rcd 8863 (2011) ($3,000,000); Main Street Telephone Co., Notice of Apparent Liability for Forfeiture, 26 FCC Rcd 8853 (2011) ($4,200,000); Norristown Telephone Co., LLC, Notice of Apparent Liability for Forfeiture, 26 FCC Rcd 8844 (2011) ($1,500,000); VoiceNet Telephone, LLC, Notice of Apparent Liability for Forfeiture, 26 FCC Rcd 8874 (2011) ($3,000,000).

[15] Assist 123, LLC, Order, 2014 WL 3512917 (Enf. Bur. 2014).

[16] Optic Internet Protocol, Inc., Notice of Apparent Liability for Forfeiture, 2014 WL 3427582 (rel. July 14, 2014) ($7,620,000): Central Telecom Long Distance, Inc., Notice of Apparent Liability for Forfeiture, 2014 WL 1778549 (rel. May 5, 2014) ($3,960,000); U.S. Telecom Long Distance, Inc., Notice of Apparent Liability for Forfeiture, 29 FCC Rcd 823 (2014) ($5,230,000); Advantage Telecommunications Corp., Notice of Apparent Liability for Forfeiture, 28 FCC Rcd 6843 (2013) ($7,600,000); Consumer Telecom, Inc., Notice of Apparent Liability for Forfeiture, 28 FCC Rcd 17196 (2013) ($3,560,000).

[17] See, e.g., Advantage Telecommunications Corp NAL., supra note 16.

[18] AT&T, Order & Consent Decree, 27 FCC Rcd 13492 (Enf. Bur. 2012); Cellco Partnership d/b/a Verizon Wireless, Order & Consent Decree, 25 FCC Rcd 15105 (Enf. Bur. 2010).

[19] See FCC Cramming Order, supra note 2.

[20] 47 C.F.R. § 64.2400-2401.

[21] Federal Communications Commission, Cramming: Unauthorized, Misleading, or Deceptive Charges Placed on Your Telephone Bill (last visited July 24, 2014), available at http://www.fcc.gov/guides/cramming-unauthorizedmisleading-or-deceptive-charges-placed-your-telephone-bill; Federal Communications Commission, Cramming Tip Sheet for Consumers (last visited July 24, 2014), available at http://www.fcc.gov/encyclopedia/cramming-tipsheet-consumers.

In: Cramming and Third Party Billing … ISBN: 978-1-63463-121-1
Editor: Christopher Rye

Chapter 7

TESTIMONY OF MICHAEL F. ALTSCHUL, SENIOR VICE PRESIDENT AND GENERAL COUNSEL, CTIA - THE WIRELESS ASSOCIATION. HEARING ON "CRAMMING ON WIRELESS PHONE BILLS: A REVIEW OF CONSUMER PROTECTION PRACTICES AND GAPS"*

On behalf of CTIA – The Wireless Association®, thank you for the invitation to participate in today's hearing. At the outset, I want to be clear – CTIA and its members share the Committee's concern about cramming. Placing an unauthorized, misleading, or deceptive third party charge on a consumer's wireless bill is wrong and simply not acceptable. That's why, in November of 2013, wireless carriers ended their support of Premium SMS services, except for charitable and political giving and inmate calling services. Moreover, carriers allow customers to block all third-party charges and have worked to make it easier for consumers to obtain refunds for unauthorized or fraudulent charges.

CTIA became involved in the industry's efforts to police Premium SMS and the associated carrier billing for these services through the Association's role as the Common Short Code Administrator. Common Short Codes allow

* This is an edited, reformatted and augmented version of testimony presented July 30, 2014 before the Senate Committee on Commerce, Science, and Transportation.

mobile users to engage and interact with a brand or service by using a short five-digit address to send text messages to a mobile application. CTIA, as the Common Short Code Administrator, assigns Common Short Codes to applicants allowing a single code to be used for the same application across multiple wireless service providers. Common short codes are used by commercial entities ranging from Dunkin Donuts to Wal-Mart, as well as by non-commercial entities, including government, charities, and political campaigns.

Short code campaigns can be employed to provide life saving information. For example, the Federal Emergency Management Agency's text message program offers regular safety tips for specific disaster types and allows for a search to find the nearest shelters and disaster recovery centers by texting "43362" ("4FEMA").[1] Another highly successful program is the Text4Baby campaign that has leveraged the power of mobile technology to help more than 700,000 new mothers and expectant women keep themselves and their babies healthy since the program's creation.[2] Not all short code campaigns are so serious -- they also can be used for purposes as varied as voting for one's favorite player or summoning an usher during a Major League baseball game.

While the overwhelming majority of short code campaigns involve no charge other than the carrier's standard rate for SMS messages, short codes also can enable users to support charities or political candidates. For example, the American Red Cross employed common short code 90999 to raise money for disaster relief in the wake of the Haitian earthquake in 2010, with donors contributing more than $43 million. A 2012 study by the Pew Internet and American Life Project found that 1 in 10 Americans has made a charitable donation through a text message.[3] Similarly, after the Federal Elections Commission granted limited approval for federal candidates, political committees, and political parties to collect political contributions through text message campaigns, the Obama for America and the Romney campaigns began using common short codes to solicit small dollar donations via mobile devices.[4] Each of these programs was, and remains, opt-in for consumers. Short codes have also been employed by state Departments of Correction to enable collect calls placed by inmates to be completed and billed to the mobile phone of the called party,[5] often a family member, who may be among the nearly 40 percent of American adults who have chosen to go "wireless-only" and forego subscribing to a wireline telephone.[6]

From their start, commercial, charitable, and political uses of common short codes issued by CTIA have been subject to written, publicly available guidelines administered both by the Mobile Marketing Association[7] and

CTIA.[8] Carriers have looked to the "connection aggregators" who link content providers to wireless carriers to supervise and enforce the consumer best practices and carrier-specific practices. In 2008, a year in which the Federal Communications Commission received only about 345 wireless cramming complaints from the nation's then 270 million wireless customers,[9] the wireless industry, both individually and through CTIA, began independent monitoring of all short code campaigns to detect any violations of the consumer best practices.

Over the next few years, the number of consumer wireless complaints filed with both the Federal Communications Commission and the Federal Trade Commission remained low and continued to decrease even with the significant growth in the number of wireless connections. While neither the FCC[10] nor FTC[11] complaint data suggested there was a significant problem in this area, the industry recognized that both wireless customers and their carriers were being victimized by fraudsters who crafted elaborate schemes to defeat the industry's self-regulation and third-partying monitoring. Where problems were alleged or identified, CTIA and its member companies worked with law enforcement officials to identify solutions and shut down entities that pursued cramming schemes.[12]

As a further safeguard to consumers, beginning in February 2012, CTIA contracted with an outside vendor to verify information supplied to the Common Short Code Administration registry by companies seeking to lease short codes for premium SMS campaigns. CTIA's vendor uses numerous commercially-available databases such as Lexis/Nexis, Dun & Bradstreet, and the Better Business Bureau to confirm that the Content Provider displayed in the registry represents a legitimate company and is identified correctly in the registry. This vetting service satisfies the requirement of the California Public Utilities Commission that carriers be responsible for the content of their bills. Any discrepancies discovered during the vetting process are communicated directly to the carriers. Although the national carriers chose to discontinue support for premium short code campaigns in late 2013,[13] CTIA has expanded the scope of its vetting to include all common short code lessees.

In addition to this front-end vetting, CTIA continues to work with outside vendors to ensure that codes are used in compliance with the applicable guidelines. When monitoring identifies problems, that information is sent to the carriers so they may take corrective action. These efforts and the national carriers' decisions to end support for premium short code campaigns (with the limited exceptions for charities and political campaigns) should combine to significantly reduce the opportunity for third-parties to use carrier billing platforms as a tool to commit fraud.

With the carriers' decisions to no longer support premium short codes for commercial campaigns, the remaining opportunities for third-party charges to appear on wireless bills are limited to instances involving direct carrier billing. Although CTIA has no first-hand knowledge of carrier practices in this area, it is our understanding that each of the national carriers employs stringent vetting and safeguards to guard against any abuse of the process.

Thank you for the opportunity to participate in today's discussion.

End Notes

[1] http://www.fema.gov/text-messages

[2] https://www.text4baby.org/.

[3] Pew Internet and American Life Project, REAL TIME CHARITABLE GIVING, (Jan.12, 2012), available at http://pewinternet.org/~/media//Files/Reports/2012/Real%20Time% 20 Charitable%20Giving.pdf at 2.

[4] http://www.huffingtonpost.com/2012/08/23/obama-text-message-donationsn1824250.html and http://www.newsmax.com/Politics/Romney-text-messages-donations/2012/08/31/id/ 4505 34/.

[5] http://www.txtcollect.com/.

[6] http://www.cdc.gov/nchs/data/nhis/earlyrelease/wireless201407.pdf at 2.

[7] http://www.mmaglobal.com/files/bestpractices.pdf.

[8] http://www.ctia.org/docs/default-source/default-document-library/industry-best-practices.pdf?sfvrsn=0, http://www.ctia.org/docs/default-source/default-document-library/guidelines-for-mobile-giving-via-wirelesscarrier-s-bill.pdf?sfvrsn=0 and http://www.ctia.org/docs/default-source/default-document-library/guidelines-forfederal-political-campaign-contributions-via-wireless-carrier-s-bill.pdf?sfvrsn=0.

[9] Empowering Consumers to Prevent and Detect Billing for Unauthorized Charges ("Cramming"), Report and Order and Further Notice of Proposed Rulemaking 27 FCC Rcd 4436 (rel. April 27, 2012)(at para. 20).

[10] Id. at para. 20-21. The last time the FCC directly reported wireless-related cramming figures was in 2002 – at which time there were 92 complaints. Since then, wireless-related cramming complaints were too few to be included in the FCC's quarterly reports on Informal Complaints and Inquiries.

[11] The FTC's 2013 Consumer Sentinel report lists the number of complaints about "Mobile: Unauthorized Charges or Debits" for 2011, 2012, and 2013 as 626, 714, and 363. See http://www.ftc.gov/system/files/documents/reports/consumer-sentinel-network-data-book-january-december2013/sentinel-cy2013.pdf at 84. To put those numbers in context, in 2011 there were 306,300,207 active subscriber units, so complaints averaged 2.0 per million subscribers. In 2012, the wireless industry served 321,716,905 active subscriber units, with complaints amounting to 2.2 per million subscribers. In 2013, subscribership rose to 326,914,000 active subscriber units, meaning complaints represented 1.1 per million subscribers, or just 0.0001% and a decline of 49% from 2012.

[12] See, for example, the statement by the Texas Attorney General, https://www.texas attorney general.gov/oagnews/release.php?id=4576, and the cases brought against Mobile Messenger, https://www.texasattorneygeneral.gov/newspubs/releases/2013/Mobile-Messen ger-POP.pdf, and Eye Level Holdings, https://www.texas attorneygeneral.gov/ newspubs/releases/2011/030911eyelevelholdings pop.pdf.

[13] Ina Fried, "AT&T, Sprint, T-Mobile, Verizon Dropping Most Premium Text Service Billing in Effort to Combat Fraud," All Things D (Nov. 21, 2013), available at http:// allthingsd.com/20131121/att-sprint-t-mobile-verizon-alldropping-most-premium-text-service-billing-in-effort-to-combat-fraud/#ina-ethics and http://www.atg. state.vt.us/news/ att-mobility-sprint-and-t-mobile-will-stop-billing-problematic-third-partycharges.php.

In: Cramming and Third Party Billing ...
Editor: Christopher Rye
ISBN: 978-1-63463-121-1

Chapter 8

CRAMMING: UNAUTHORIZED, MISLEADING, OR DECEPTIVE CHARGES PLACED ON YOUR TELEPHONE BILL*

Federal Communications Commission

BACKGROUND

"*Cramming*" is the practice of placing unauthorized, misleading or deceptive charges on your telephone bill. Crammers rely on confusing telephone bills in an attempt to trick consumers into paying for services they did not authorize or receive, or that cost more than the consumer was led to believe.

HOW DOES CRAMMING OCCUR?

In addition to providing local telephone service, local telephone companies often bill their customers for long distance and other services that other companies provide. If a local telephone company, long distance telephone company or another type of service provider either accidentally or

* This is an edited, reformatted and augmented version of a consumer guidance document, issued January 2013.

intentionally places unauthorized, misleading or deceptive charges on your bill, you may have been "crammed."

Cramming can also occur if a local or long distance company or another type of service provider does not clearly or accurately describe all of the relevant charges to you when marketing a service. Although you may have authorized the service, you did not understand or were misled about how much it would really cost.

CRAMMING CHARGES: WHAT THEY LOOK LIKE

Cramming comes in many forms and is often hard to detect unless you closely review your telephone bill. Charges described as follows can be legitimate if you authorize them but, if unauthorized, could constitute cramming:

- Charges for services that are explained on your telephone bill in general terms such as "service fee," "service charge," "other fees," "voicemail," "mail server," "calling plan," "psychic" and membership;"
- Charges that are added to your telephone bill every month without a clear explanation of the services provided – such as a "monthly fee" or "minimum monthly usage fee;" and
- Charges for an authorized service, but you were misled about its actual cost.

While cramming charges typically appear on consumers' local telephone bills, they may also be included with bills issued by long distance telephone companies and companies providing other types of services, including wireless telephone, beeper and pager services.

THE FCC'S TRUTH-IN-BILLING RULES

The Federal Communications Commission's (FCC) Truth-in-Billing rules require telephone companies to provide clear, non-misleading, plain language in describing services for which you are being billed.

Because one telephone company, usually your local telephone company, may include charges you incurred for another company's service on your bill, the company sending you the bill must identify the service provider associated with each charge. If a bill contains charges in addition to basic local service, it must distinguish between charges for which non-payment will result in disconnection of basic, local service and charges for which non-payment will not result in disconnection. Telephone companies must also display, on each bill, one or more toll-free numbers that you can call to ask about or dispute any charge on the bill. This information empowers you to protect yourself from cramming and other types of telecommunications fraud. It also helps you make informed choices when you shop around to find the best telephone service to meet your needs.

How to Protect Yourself and Save Money

- Carefully review your telephone bill every month. Treat your telephone service like any other major consumer purchase or service. Review your monthly bills just as closely as you review your monthly credit card and bank statements.
- Ask yourself the following questions as you review your telephone bill:
    1. Do I recognize the names of all the companies listed on my bill?
    2. What services were provided by the listed companies?
    3. Does my bill include charges for calls I did not place and services I did not authorize?
    4. Are the rates and line items consistent with the rates and line items that the company quoted to me?
- You may be billed for a call you placed or a service you used, but the description listed on your telephone bill for the call or service may be unclear. If you don't know what service was provided for a charge listed on your bill, ask the company that billed the charge to explain the service provided before paying the charge.
- Make sure you know what service was provided, even for small charges. Crammers often try to go undetected by submitting $2.00 or $3.00 charges to thousands of consumers.
- Keep a record of the telephone services you have authorized and used – including calls placed to 900 numbers and other types of telephone

information services. These records can be helpful when billing descriptions are unclear.
- Carefully read all forms and promotional materials – including the fine print – before signing up for telephone or other services to be billed on your phone bill.

Companies compete for your telephone business. Use your buying power wisely and shop around. If you think that a company's charges are too high or that its services do not meet your needs, contact other companies and try to get a better deal.

ACTIONS YOU CAN TAKE IF YOU'VE BEEN CRAMMED

Take the following actions if unknown or suspicious charges are listed on your telephone bill:

- Immediately call the company that charged you for calls you did not place, or services you did not authorize or use. Ask the company to explain the charges. Request an adjustment to your bill for any incorrect charges.
- Call your own local telephone company. Explain your concerns about the charges and ask your local telephone company the procedure for removing incorrect charges from your bill.

FILING A COMPLAINT

If neither the company sending you the bill nor the company that provided the service in question will remove charges from your telephone bill that you consider to be incorrect, you can file a complaint as follows:

- with the FCC for charges related to telephone services between states or internationally,
- with your state public service commission for telephone services within your state; and
- with the FTC for non-telephone services on your telephone bill.

For charges related to telephone service between states or internationally, you can file your complaint with the FCC. There is no charge for filing a complaint. You can file a complaint using an FCC online complaint form found at_www.fcc.gov/complaints. You can also file your complaint with the FCC's Consumer Center by calling 1-888-CALL-FCC (1-888-225-5322) voice or 1-888-TELL-FCC (1-888-835- 5322) TTY; faxing 1-866-418-0232; or writing to:

Federal Communications Commission
Consumer and Governmental Affairs Bureau
Consumer Inquiries and Complaints Division
445 12th Street, SW
Washington, DC 20554.

What to Include in Your Complaint

The best way to provide all the information needed for the FCC to process your complaint is to complete fully the online complaint form. When you open the online complaint form, you will be asked a series of questions that will take you to the particular section of the form you need to complete. If you do not use the online complaint form, your complaint, at a minimum, should indicate:

- your name, address, email address and phone number where you can be reached;
- the telephone and account numbers that are the subject of your complaint;
- the names and phone numbers of any companies involved with your complaint;
- the amount of any disputed charges, whether you paid them, whether you received a refund or adjustment to your bill, the amount of any adjustment or refund you have received, an explanation if the disputed charges are related to services in addition to residential or business telephone services; and
- the details of your complaint and any additional relevant information.

Filing a Complaint with Your State Public Service Commission or the FTC

For charges for telephone services provided within your state, contact your state public service commission. For contact information for your state public service commission, visit www.naruc.org/commissions.cfm or check the blue pages or government section of your local telephone directory.

For charges on your telephone bill for non-telephone services, file your complaint with the Federal Trade Commission (FTC). You can file a complaint with the FTC online at www.ftccomplaintassistant.gov. You can also submit a complaint by calling the FTC toll-free at 1-877-382-4357 (voice) or 1-866-653-4261 (TTY), or writing to:

Federal Trade Commission
CRC-240
600 Pennsylvania Ave., NW
Washington, DC 20580.

For More Information

For information about other telecommunications issues, visit the FCC's Consumer website at www.fcc.gov/consumers, or contact the FCC's Consumer Center using the information provided for filing a complaint.

Last Reviewed: 01/18/13.

INDEX

A

B

C

D

E

F

S

T

U

V

W